Innovations in food labelling

Related titles:

EU food law
(ISBN 978-1-85573-557-6)
This book is an ideal introduction to this complex field for the non-specialist. The book opens with an introduction to the key principles of EU food law and the main mechanisms of enforcement. Subsequent chapters cover legislation in specific areas including hygiene, additives and functional foods.

Food labelling
(ISBN 978-1-85573-496-8)
J Ralph Blanchfield has brought together a distinguished panel of European and American experts from industry and academia. Together they provide a comprehensive guide to producing labelling that complies with existing legislation and satisfies the needs of all interested parties. *Food labelling* also provides a valuable insight into possible future developments.

Understanding consumers of food products
(ISBN 978-1-84569-009-0)
It is very important for food businesses, scientists and policy makers to understand consumers of food products: in the case of businesses to develop successful products and in the case of policy makers to gain and retain consumer confidence. Consumers' requirements and desires are affected by issues such as culture, age and gender, and issues important to consumers nowadays such as diet and health or GM foods will not always be so significant. Therefore food businesses and policy makers need to understand consumers' attitudes and the influences upon them to respond effectively. Edited by two distinguished experts, this book is an essential guide for food businesses, food scientists and policy makers.

Details of these books and a complete list of Woodhead's titles can be obtained by:

- visiting our web site at www.woodheadpublishing.com
- contacting Customer Services (e-mail: sales@woodheadpublishing.com; fax: +44 (0) 1223 893694; tel.: +44 (0) 1223 891358 ext. 130; address: Woodhead Publishing Limited, Abington Hall, Granta Park, Great Abington, Cambridge CB21 6AH, UK)

Innovations in food labelling

Edited by
Janice Albert

Published by
The Food and Agriculture Organization
of the United Nations
and
Woodhead Publishing Limited

CRC Press
Boca Raton Boston New York Washington, DC

WOODHEAD PUBLISHING LIMITED
Oxford Cambridge New Delhi

Published by the Food and Agriculture Organization of the United Nations and Woodhead Publishing Limited.

Woodhead Publishing Limited, Abington Hall, Granta Park, Great Abington, Cambridge CB21 6AH, UK
www.woodheadpublishing.com

Woodhead Publishing India Private Limited, G-2, Vardaan House, 7/28 Ansari Road, Daryaganj, New Delhi – 110002, India
www.woodheadpublishingindia.com

Published in North America by CRC Press LLC, 6000 Broken Sound Parkway, NW, Suite 300, Boca Raton, FL 33487, USA

First published 2010, Woodhead Publishing Limited and CRC Press LLC

British Library Cataloguing in Publication Data
A catalogue record for this book is available from the British Library.

Library of Congress Cataloging in Publication Data
A catalog record for this book is available from the Library of Congress.

Woodhead Publishing ISBN 978-1-84569-676-4 (book)
Woodhead Publishing ISBN 978-1-84569-759-4 (e-book)
CRC Press ISBN 978-1-4398-2484-9
CRC Press order number: N10137
FAO ISBN 978-92-5-106180-0

The publishers' policy is to use permanent paper from mills that operate a sustainable forestry policy, and which has been manufactured from pulp which is processed using acid-free and elemental chlorine-free practices. Furthermore, the publishers ensure that the text paper and cover board used have met acceptable environmental accreditation standards.

Typeset by Ann Buchan (Typesetters), Middlesex, UK
Printed by TJ International Limited, Padstow, Cornwall, UK

Front cover image © FAO/Roberto Faidutti

Contents

Foreword

Food producers in many countries are keenly interested in finding ways to inform consumers about the qualities of their products at the point of purchase. Many consumers actively seek information about products that have qualities that serve their health needs and are consistent with their values. As a result of these varied interests, food labels are increasingly being used to provide consumers with information about the environmental, technical and socioeconomic conditions under which the products were produced, as well as the health and safety aspects of food products.

The growing consumer and industry interest in food labels presents challenges for government authorities, which must ensure that the information that appears on food packages is useful, credible and presented clearly so that it does not mislead the consumer. With the increase in global trade in food, there is a need to harmonize food labelling so that product information is easily understood and is relevant to consumers in different markets.

Developing and implementing food labelling policies is a complex undertaking that presents many challenges. This book illustrates the multiple purposes food labelling serves and the many steps that different actors must take to implement a successful labelling policy.

FAO, with its breadth of technical expertise and practical experiences in many areas of food production, nutrition, food safety, marketing and trade as well as social development, and its key role in developing global food standards, food laws and international treaties related to food, agriculture, fisheries and forestry, is uniquely positioned to assist governments, food producers and consumers in understanding the dynamic role of food labelling in the global food system today.

FAO is pleased to collaborate in producing this publication with Woodhead Publishing Limited, which specializes in scientific and technical advice related to food. We would like to express our appreciation to the authors who have generously shared their expertise and experiences in this effort. In addition, we would like to thank a number of colleagues who provided important technical advice and administrative assistance in producing this book: Rachel Tucker, Selma Doyran, Nadia Scialabba, Annika Wennberg, Ellen Muehlhoff, Maria Lourdes Costarrica, Valeria Menza, David Hallam, Jayne Beaney, Johanna Lyons, Mercedes Casellas and Katherine Emery.

Ezzeddine Boutrif
Director
Nutrition and Consumer Protection Division
Food and Agriculture Organization of the United Nations
Rome, Italy

1

Introduction to innovations in food labelling

Janice Albert, Food and Agricultural Organization of the United Nations (FAO), Italy

Abstract: National labelling laws, international norms and guidelines as well as private standards aim to protect consumers from deception and businesses from unfair competition. Food labelling is also becoming a policy tool for motivating change in consumer behaviour and shifts in food production practices. In this process of developing labels, the interactions between private actors and public institutions are dynamic and complex, especially given the need to harmonize labels to facilitate trade. This book provides information about the rights and responsibilities that are the foundation for food labelling, and illustrates how labelling policies are developed. Labelling topics include the Codex Committee on Food Labelling, international trade agreements and human rights, nutrition, allergens, organic, eco-labelling for fish, fair trade, geographic indication and genetically modified foods.

Key words: labelling principles, international food standards, labelling and health, labelling and environment, labelling and socioeconomic conditions.

1.1 Introduction: the evolution of food labelling

In their broadest and most conventional application, food labelling policies have a dual purpose: to protect consumers and to ensure fair marketing. National laws, international norms and guidelines as well as private standards prohibit labelling that misrepresents the quality of a product and deceives consumers. Prevention of false claims protects businesses from unfair competition. These basic principles were established long ago and they are still highly relevant.

All consumers need to be able to rely upon the truthfulness of information on a package, which helps them to distinguish among products and to make proper use of the products. At best, labels are part of the environment that enables consumers to make food choices according to their needs and desires. For labels to serve their intended purposes they should be accompanied by education and information. Yet, all too often, these resources are not available. This is why it is essential that labels are easy to understand and that those responsible for food package information do not take advantage of vulnerabilities.

For some, food labelling is seen as more than a form of minimal protection; it is a policy tool for motivating change in consumer behaviour and different food production practices. Increasingly, labelling relates specific products to consumers' interests in health, the environment, culture and social well-being. As labelling policies encompass a larger number of topics, there are more interactions between private actors and public institutions. Each decision reflects a particular regulatory approach and state of knowledge, with the influences of different actors, agencies and events varying greatly. This leads to inconsistent approaches regarding labels and differences of opinion, even among experts. Within the same country or organization, one can find labelling policies that are very restrictive and others that are very permissive in terms of the type of information that may be placed on a package.

When considering international standards, the situation becomes even more intricate as each national organization brings its labelling ideas into the international arena and the process of harmonization begins. As more food is traded and labelling must meet the needs of consumers in different countries, the process for establishing specific labelling standards has become very challenging. The implications of every detail of a label are scrutinized before governments reach consensus on a standard, and the label gains acceptance of stakeholders. There is large scope for interpretation of the basic principles for specific foods and markets and a trend towards expansion of the principles, which can lead to contentious, lengthy and costly debates.

To facilitate the development and use of food labelling, more understanding of good labelling practices is needed among governments, industry, civil society organizations and consumers. This book aims to contribute to this goal by providing essential information about the rights and responsibilities that are the foundation for food labelling, and providing case studies of labels that are currently at different stages of development. Collectively, the chapters in this book provide a rich picture of the dynamic and multi-faceted topic of food labelling. While the subject of each chapter is different, there are common features and processes that can be discerned.

1.2 Standards and legal issues

Every food label must comply with food laws and standards. Because of the increased importance of the food trade, national authorities often pay close attention to harmonizing their laws with international standards. The Codex Alimentarius Commission is the recognized international authority for food standards setting. Since the Food and Agriculture Organization and the World Health Organization established the programme in 1962, labelling has been a fundamental aspect of the work. In fact, the first Codex food standard was a labelling standard. In Chapter 2, Randell explains the major labelling standards developed by Codex and how the work of the Codex Committee on Food Labelling is evolving. The Codex standards are increasingly relevant, particularly because they are recognized in international trade agreements of the World Trade Organization. In Chapter 3, Vidar explains the key international trade agreements and relates them to the international commitments and treaties to protect the rights of consumers. She notes that consumers have a right to affordable foods, thus it is important to consider the costs as well as the benefits of labelling.

1.3 Labelling to protect and promote health

In the 21st century, the information that is considered to be necessary or desirable for consumers to protect their health is changing dramatically. With the accumulation of scientific evidence linking food and health, there is a trend towards voluntary and mandatory food labelling as a tool to address nutrition-related problems. The latest trends in nutrition labelling are discussed by Hawkes in Chapter 4. As new foods become available in different markets, additional measures are being taken to protect consumers who may be allergic to certain foods, yet unaware that the foods are ingredients in unfamiliar products. These consumers need labels to warn them since they cannot detect which products contain ingredients that cause allergic reactions on their own. In Chapter 5, Hattersley and Chun-Han explain how food authorities ensure that the necessary information is available, without causing unnecessary dietary restrictions.

1.4 Labelling to protect the environment and promote sustainable food production

Labelling contributes to the efficient functioning of the market by enabling consumers to express their preferences, which may be based on values and interest as well as tastes, budgets and health. Public concern about the impact of food production on the environment has stimulated interest in labelling of organic foods and eco-labelling. Environmental organizations have promoted the use of food labels as a strategy of providing market incentives to encourage more sustainable production practices. The aims of the organic food industry and public and private procedures for protecting the integrity of their product labels are discussed by Compagnoni in Chapter 6. In Chapter 7, Willmann, Cochrane and Emerson explain the need for sustainable marine fishing practices and how eco-labelling may motivate better practices. The latest information on an international code within this industry is described and the process of developing the code is reviewed.

1.5 Labelling to promote social well-being and protect culture

Consumers may express their interest in preservation of traditional cultures and specific foods through their purchasing decisions. Others wish to support food producers in developing countries and food production that provides decent economic and social conditions. Labelling is a means for food producers to inform consumers about their ways of producing foods. With markets for products being thousands of miles from the place of production, consumers cannot determine whether the claims about production practices are true without certification by independent sources. Through certification, consumers gain confidence in the truthfulness of a specific label. In Chapter 7, Liu describes the business case for certification in relation to fair trade products. In Chapter 8, Vandecandelaere raises our awareness of the social dimension in food labelling in discussing geographic indicators (GI). GI labels have been used for centuries to distinguish foods with unique qualities; today they are being used to raise esteem and earnings for local producers. GI labels are intended to recognize and protect producer's rights and protect culture and traditions.

1.6 Labelling in relation to new technologies

Governments and food producers need to stimulate economic growth and innovation, which may occur with new technologies for food production. At the same time, they must respect the views of citizens who may not be in favour of particular technological changes. Labelling is often preferred as a policy tool in such situations because it does not restrict a product from being marketed but it allows consumers to express their views through their purchases. In theory, the market will determine whether a technology will succeed since labelling provides information to buyers and their actions give a signal to sellers about consumer preferences. As shown in the examples above, the food producer expects to be rewarded for practices that are desired by consumers. Labelling can also enable consumers to reject a product, with the loss in sales causing the producer to remove the product from the market.

In the case of genetically modified foods, labelling has been proposed as a way to allow consumers to demonstrate their views about the technology. In Chapter 10, Albert reports on the seven major labelling options for GM foods found among the countries that belong to the Codex Alimentarius Commission. She provides information about two of the most controversial approaches, the voluntary approach of the United States of America and the mandatory approach of the European Union.

2

The Codex Alimentarius and Food Labelling: delivering consumer protection

Alan W. Randell, former Secretary of the Codex Alimentarius Commission, Italy

Abstract: The Codex General Standard for the Labelling of Prepackaged Foods and other Codex texts dealing with claims in general and health and nutrition claims in particular are described in a historical context. It is possible to demonstrate a transition from trade-based standards to consumer-based standards and guidelines – a process that is still underway. The main shift is from a strict 'prevention of fraud' scenario (1960s) to a scenario of providing consumer information (1990s) and, finally, to a scenario of delivering health policy through labelling.

Key words: Codex Alimentarius, food labelling, consumer protection, fraudulent claims, nutrition, health claims, 'organic' food, irradiated food, 'Halal' food, biotechnology, allergens, diet, non-communicable disease.

2.1 Introduction

The Codex General Standard for the Labelling of Pre-Packaged Foods was adopted in 1969: the first international standard to be approved by the newly-formed Codex Alimentarius Commission (CAC, 1969). The fact that the Standard bears the number "Codex-Stan 1" reflects its central importance as the principal Codex standard for consumer protection and for ensuring fair practices in the food trade. The General Standard was extensively revised and enlarged in 1985 and, since then, numerous amendments and additions have ensured that the Standard remains the key Codex instrument for delivering information about food to the consumer.

The General Standard is accompanied by a range of specific texts. All Codex commodity standards provide specific interpretation of the General standard in reference to the 'Name of the Food', and two additional standards cover the labelling of foods for special dietary uses and of food additives when they are sold as such to the consumer and there are guidelines on nutrition labelling. Additional interpretation of the Standard is provided by guidelines covering the legitimacy of claims made on food labels: claims in general; health and nutrition claims; and claims concerning production and processing methods associated with 'organic' foods and 'Halal' foods.

Current debate about food labelling in Codex concerns the degree to which labelling should be used an instrument of delivering public policy, in particular nutrition policies related to diet, physical activity and chronic diseases. The question of how to label GM foods remains unsolved.

This chapter will review the evolution of food labelling in Codex and the change in emphasis from trade-related standards to consumer-based standards and as an instrument in delivering public health policy.

2.2 The Codex framework for food labelling

The Codex framework for food labelling consists of the General Standard, specific provisions in Codex commodity standards, a series of interpretative guidelines dealing with types of claims, and standards for the labelling of special dietary and special medical foods and for food additives sold as such. A specific standard deals with terms to be used for the labelling of dairy products (see Box 2.1).

The body responsible for the preparation of labelling texts is the Codex Committee on Food Labelling, hosted by the government of Canada since the inception of the Commission's work in this area. This Committee receives advice from other Codex Committees, most notably the Committee on Nutrition and Foods for Special Dietary Uses on matters relating to the technical and scientific basis of provisions concerning nutrition and health labelling, and the Codex Committee on Food Additives on issues relating to the names used for food additives.

The General Standard applies to all prepackaged foods offered to the consumer and for catering purposes. It is supplemented by individual Codex commodity standards that describe the nature and composition of foodstuffs and specify the name and/or names of the food reserved for use on products that conform to these standards. To a certain extent therefore, all Codex commodity standards are extensions of the General Standard and assist in its interpretation. Exemptions from, or additions to, the General Standard are allowed if they are necessary for interpretation in respect of the product concerned, including the provisions concerning date-marking (CAC, 2008b). In 2002, the combination of the General Standard and the specific Codex Standard for Sardines and Sardine-like Products

Box 2.1 Codex texts on food labelling

General Standard for the Labelling of Prepackaged Foods (Codex Stan 1-1985)
General Standard for the Labelling of Food Additives when Sold as Such (Codex Stan 107-1981)
General Standard for the Labelling of and Claims for Prepackaged Foods for Special Dietary Uses (Codex Stan 146-1985)
Codex Standard for the Labelling of and Claims for Foods for Special Medical Purposes (Codex Stan 180-1991)
General Standard for the Use of Dairy Terms (Codex Stan 206-1999)
General Guidelines on Claims (CAC/GL 1-1979)
Guidelines for Nutrition Labelling (CAC/GL 2-1985)
Guidelines for the Use of Health and Nutrition Claims (CAC/GL 23-1997)
General Guidelines for the Use of the Term 'Halal' (CAC/GL 24-1997)
Guidelines for the Production, Processing, Marketing and Labelling of Organically Produced Foods (CAC/GL 32-1999)

(Codex Stan 94-1981, Rev.1-1995) was used by the World Trade Organization for the resolution of a dispute under the provisions of the Agreement on Technical Barriers to Trade (WTO, 2002).

The Committee on Food Labelling reviews and endorses the labelling provisions of all draft standards submitted to the Commission for adoption to ensure consistency with the General Standard. In recent years, there has been a trend towards full harmonization of requirements with those of the General Standard and a reduction in the number of exemptions or special requirements.

2.3 Specific food labelling issues in the Codex general standard

2.3.1 The principles of food labelling

The first substantive section of the General Standard contains the 'Principles of Food Labelling' (Box 2.2). These principles have not been modified since the original 1969 standard. They strongly reflect that particular part of the mandate of the Codex Alimentarius Commission aimed at 'ensuring fair practices in the food trade' but do not address the other main objective of the Commission, namely protecting the health of consumers. There have been amendments to the General Standard with this latter objective in mind, in particular in relation to the labelling of potential allergens, and current trends, particularly the use of labelling for the promotion of public nutrition policies, are more closely linked to this objective.

It may also be questioned whether the Codex principles are sufficient to meet the legitimate need of consumers 'to have access to adequate information to enable them to make informed choices according to individual wishes and need' as set out in the United Nations Guidelines on Consumer Protection (United Nations, 1985).

Box 2.2 The principles of food labelling

Prepackaged food shall not be described or presented on any label or in any labelling in a manner that is false, misleading or deceptive or is likely to create an erroneous impression regarding its character in any respect;
Prepackaged food shall not be described or presented on any label or in any labelling by words, pictorial or other devices which refer to or are suggestive either directly or indirectly, of any other product with which such food might be confused, or in such a manner as to lead the purchaser or consumer to suppose that the food is connected with such other product.

Various actions taken by the Commission have extended the informational aspect of food labelling, in particular in date-marking and in nutrition labelling, with a view to enabling consumers to make informed choices.

2.3.2 Ingredient labelling

The labelling of food ingredients, especially food additives, has been a feature of the Codex General Standard since its inception, and this is the most detailed section of the Standard. Such labelling has consumer information as its primary objective as it is a general principle that any substance added to food as an additive must have been evaluated and approved for safety in use. However, should a consumer wish to avoid any particular additive, this section of the standard allows such a choice to be made. Ingredient labelling has been the subject of multiple amendments since the adoption of the original standard; all tending towards greater transparency in the information provided to the consumer in particular in the declaration of sub-ingredients in combined foods or mixtures of foods and in the quantitative labelling of certain ingredients (CCFL, 2008a). There are specific requirements for the declaration of potential allergens (see below).

2.3.3 Date-marking

Detailed date-marking provisions were included in the General Standard at the time of its revision in 1985. Prior to this revision, date-marking was considered on a case-by-case basis by Codex commodity committees on the basis of Guidelines developed by the Committee on Food Labelling and adopted by the Commission in 1978 (revised in 1981). The Codex General Standard introduced the use of the expression ‘Best before’ to describe the date of minimum durability (the preferred form of date-marking), and this expression is now widely used in national regulation. Although not in conflict with the Principles of Food Labelling, the provision of information about expected quality by means of date-marking extends the amount of information provided to the consumer through food labelling.

There remains some confusion as to the purpose of date-marking, in particular in reference to food safety, and unfortunately the General Standard is not explicit on this. On the other hand, the Guidelines to Codex Committees were quite explicit and stated that: "The purpose of date marking is to give the consumer a date which will provide information about the expected quality of the product provided that it has been properly stored. This does not mean that date marking guarantees either the acceptability or the safety of the product" (CCFL, 1980). Date-marking should be accompanied by appropriate storage instructions taking into account the nature of the product.

2.3.4 Irradiated food

The General Standard contains a brief, but detailed, section on the labelling of irradiated food. This is unusual as firstly, such requirements would normally have been included in the corresponding commodity standard, in this case the General Standard for Irradiated Foods (Codex Stan 106-1983, Rev.1-2003); and secondly, because this is the only processing and production method singled out for special labelling. The historical reasons for this special treatment in part have to do with the consideration of the use of food irradiation as having the same characteristics as the use of food additives (technical discussions were in fact handled by the Codex Committee on Food Additives) and in part because of the heightened interest of consumers in this matter. The inclusion of provisions relating to specific processing and production methods has also been considered for the labelling of foods obtained through certain techniques of genetic modification/genetic engineering but without agreement on this matter[1] (see below).

2.4 Claims and other interpretative guidelines

Following the adoption of the General Standard in 1985, there was considerable debate in the Committee on Food Labelling as to the nature of claims that might be made legitimately within the context of the second of the Principles of Food Labelling. Among those claims that caused concern were exaggerated health and nutrition claims, claims that a normal diet could not provide adequate nutrition, and claims concerning the use of specific terms such as 'natural' or 'organic' (CCFL, 1972). The debate led to the adoption of the General Guidelines on Claims to serve as an amplification of the second statement of principles (CCFL, 1979). The Guidelines were subsequently revised in 1991.

The General Guideline on Claims deals with prohibited claims, potentially misleading claims and conditional claims (i.e., claims subject to specific conditions).

Specific texts have since been adopted by the Commission to provide further detailed interpretation in the case of foods marketed as 'organically-produced' and foods claimed to be 'Halal'. Claims made in relation to special dietary foods and medical foods are covered by separate standards (see Box 2.1).

---

[1]At 1 January 2009.

The Committee considered, but then abandoned, draft guidelines on the use the term 'natural' (CCFL, 1994) and 'vegetarian' (CCFL, 2000). As of 2009, the Committee has not undertaken work on social or environmental claims (other than 'organic') nor on claims related to geographic identification.

2.5 Nutrition labelling: health and nutrition claims

The Commission has adopted two major texts in this area: Guidelines for Nutrition Labelling and Guidelines for Use of Health and Nutrition Claims (see Box 2.2). Both texts originated from discussions about how nutrition and health information should be conveyed to the consumer.

The current Guidelines on Nutrition Labelling were adopted in 1985 and have been amended several times; most recently in 2006. They describe the form and content of nutrition information on a food label. This information can be applied voluntarily, but must be applied when either a nutrition claim or a health claim is made. The Guidelines for the Use of Health and Nutrition Claims were adopted in two phases: the first dealing with nutrition claims only (1997) and the second dealing with both health and nutrition claims (2004). Both sets of guidelines were developed with the technical and scientific advice of the Codex Committee on Nutrition and Foods for Special Dietary Uses.

Although both of these guidelines contain provisions that expand and interpret the General Standard, they add very substantially to the Principles of Food Labelling which address almost exclusively the prevention of fraud or deceptive practices.

The Guidelines on Nutrition Labelling introduced the concept of consumer information 'so that a wise choice … can be made', and linked nutrition labelling to public health policy (see Box 2.3). The preamble to the Guidelines for Use of

Box 2.3 Purpose of the guidelines on nutrition labelling

To ensure that nutrition labelling is effective:
in providing the consumer with information about a food so that a wise choice of food can be made;
in providing a means for conveying information of the nutrient content of a food on the label;
in encouraging the use of sound nutrition principles in the formulation of foods which would benefit public health;
in providing the opportunity to include supplementary nutrition information on the label.
To ensure that nutrition labelling does not describe a product or present information about it which is in any way false, misleading, deceptive or insignificant in any manner.
To ensure that no nutritional claims are made without nutrition labelling.

Nutrition and Health Claims also linked labelling to the implementation of national policies to the point that 'only nutrition claims that support national nutrition policy should be allowed' and that health claims should be consistent with, and support, national health policies, where applicable (CAC, 2008a). These changes should be seen as a shift of emphasis from 'ensuring fair practices in the food trade' to a more balanced approach that also encompasses 'protecting the health of consumers'. Both objectives are equally stressed in the Statutes of the Codex Alimentarius Commission. The provision of information (almost as an end in itself) is consistent with the general principles contained in the UN Guidelines for Consumer Protection that call for 'access of consumers to adequate information to enable them to make informed choices according to individual wishes and needs' (United Nations, 1985).

The Guidelines on Nutrition Labelling provide for the declaration of energy value; protein, carbohydrate and fat content; and the amount of any other nutrient for which a nutrition or health claim is made. There are specific provisions for the declaration of claims relating to types of fatty acids and cholesterol; types of carbohydrates and fibre; and for vitamins and minerals. The Guidelines also specify the way in which nutrient content should be presented on the label, including a list of reference or daily intake values for certain nutrients.[2]

The Guidelines for Use of Health and Nutrition Claims allow nutrition and nutrient claims as well as comparative nutrient claims, under specified conditions and in conformity with standardized wording for such claims. Health claims must be substantiated scientifically and must consist of information on the physiological role of the nutrient(s) or on an accepted diet–health relationship, and relevant information on the composition of the food unless the diet–health relationship is based on the whole food or cannot be linked to specific constituents of the food. Claims may be made that relate to established dietary guidelines or to 'health diets'. A table of conditions as to what constitutes 'Low', 'Very low', 'Free' and 'Source' is included in the Guidelines. An annex on the scientific substantiation of health claims was submitted to the Commission for adoption in 2009 (CCNFSDU, 2008).

2.5.1 Diet, physical activity and health

In May 2004, the WHO's World Health Assembly endorsed a global strategy to combat the increasing world-wide incidence in non-communicable diseases, in particular cardio-vascular disease, type 2 diabetes and certain types of cancer linked to unhealthy diets and physical inactivity. The strategy was developed from the report of an FAO/WHO Expert Consultation on Diet, Nutrition and the Prevention of Chronic Diseases (FAO/WHO, 2003). The global strategy envisages further development in certain areas for which the Codex Alimentarius Commission

[2] At the time of writing, the Codex Committee on Nutrition and Foods for Special Dietary Uses was considering the development of general principles for establishing nutrient reference values of vitamins and minerals for the general population (CCNFSDU, 2008).

is competent including: labelling to allow consumers to be better informed about the benefits and content of foods; measures to minimize the impact of marketing on unhealthy dietary patterns; fuller information about healthy consumption patterns, including steps to increase the consumption of fruit and vegetables; and production and processing standards regarding the nutritional quality and safety of products (WHO, 2004).

Both the Codex Committee on Food Labelling and the Codex Committee on Nutrition and Foods for Special Dietary Uses have begun work in response to the global strategy. At the time of writing,[3] the Committee on Food Labelling is discussing a revision of the Guidelines on Nutrition Labelling including:

- the list of nutrients that are always declared on a voluntary or mandatory basis and discussion of issues related to mandatory nutrition labelling;
- the legibility and readability of nutrition labelling and labelling provisions dealing with the food ingredients identified in the global strategy; and
- the revision of the Nutrient Reference Values contained in the Guidelines (CCFL, 2008b).

The Committee on Nutrition and Foods for Special Dietary Uses has decided to undertake new work on the development of Nutrient Reference Values associated with increased or decreased risk of non-communicable diseases (CCNFSDU, 2008).

The final response to the WHO Global Strategy, in the form of new or revised Codex standards or guidelines, is certain to take several years of scientific enquiry and intense negotiation within the relevant Codex committees. Nevertheless, the principle that these standards and guidelines should reflect public health and consumer protection policies has already been established. The work continues.

2.6 Labelling, food safety and allergens

It is universally acknowledged that foods offered for sale to consumers should be safe to eat, at least when prepared and/or cooked according to usual household practices. In fact, people have the right to expect the food they eat to be safe and suitable for consumption (CAC, 2003). It is not surprising, therefore, that all Codex food labelling assumes that all foods to which labelling might apply are safe to eat. Of course, foods can deteriorate and become unsuitable for human consumption, and labelling plays a significant role in limiting the effects of deterioration by providing suitable storage and handling instructions. The Codex Principles and Guidelines for the Conduct of Microbiological Risk Management (CAC/GL 63-2007) suggest that proper labelling includes information that instructs the consumer regarding safe handling practices and, where appropriate, briefly informs the consumer of the food safety issue.

For some sections of the population the problem is not microbiological risk but

[3]February 2008.

allergies and intolerances to certain foods or food ingredients that cause discomfort, illness, or even death. A detailed examination of food labelling and allergies is provided elsewhere in this book (Chapter 5). The Codex General Standard for the Labelling of Prepackaged Foods requires that certain known allergens [4] always be included in the list of ingredients even if they are present as sub-ingredients of composite foods below the cut-off level of 5 per cent. The presence of any of these allergens occurring as a result of transfer through genetic modification must also be declared.

2.7 Foods derived from biotechnology

The Codex Alimentarius Commission first discussed the implications of modern biotechnology for its food standards work in 1989 (CAC, 1989), and the matter was first taken up by the Committee on Food Labelling in 1994. Discussions within the Committee have been difficult and, at the time of writing, no texts relating to the labelling of foods derived from biotechnology ('genetically modified' or 'genetically engineered' foods) have been adopted by the Commission, with the exception of an amendment to the General Standard concerning the transfer of known food allergens into a modified food. A more complete description of the issues surrounding the labelling of foods derived from biotechnology is given in Chapter 9 of this book.

2.8 Codex, labelling and advertising

The question of advertising and its relationship to labelling has been the subject of debate within the Codex Alimentarius Commission and its Committee on Food Labelling over many years. A chronology of this debate, up to and including the Commission session in 2003, was prepared by the Canadian Secretariat to the Committee (CCFL, 2004). Since then, the Commission has included a reference to advertising in the Guideline for Use of Nutrition and Health Claims "where required by the authorities having jurisdiction" and a minor amendment in the form of a definition for advertising in the same Guideline (CAC, 2008).

In 1984 a legal opinion was provided by the Legal Counsels of FAO and WHO (cited in CCFL, 2004) that contains the following basis for the consideration of advertising by the Commission:

> 'Advertising' is not specifically referred to in the Statutes of the Codex Alimentarius Commission, whose mandate is the implementation of a program designed to protect the health of consumers and to ensure fair practices in the food trade. However, to carry out such a mandate, the Commission is implicitly authorized to deal with matters which are

[4] Cereals containing gluten; crustacean, eggs, fish, peanuts, soybeans, milk, tree nuts and the products of any of these; and sulphite in concentrations of 10 mg/kg or more.

> necessarily incidental and ancillary to the very substance of such mandate. 'Advertising' has always been considered, both generally and by the parent organizations of the Commission, as a matter having aspects which are necessarily incidental and ancillary to the protection of the health of the consumers and the ensurance of fair practices of (food) trade. The Commission may therefore consider the aspects of advertising which are directly linked to the achievement of its purpose. It may also delegate its competence relating to advertising to a subsidiary body, in accordance with its Statutes.

Despite this legal authority, the Commission has been most reluctant to enter into in-depth discussions on advertising and there are very few references to advertising in Codex standards and guidelines. These include:

- A statement to the effect that nothing in the labelling and advertising of foods for special dietary uses foods shall imply that advice from a qualified person is not needed (Codex Stan 146-1985, Section 3.2).
- A statement that the advertising of foods for special medical purposes to the general public should be prohibited (Codex Stan 180-1991, Section 3).
- A claim made in advertising that a food is 'organic' is sufficient to trigger the Guidelines for the Production, Processing, Marketing and Labelling of Organically Produced Foods (CAC/GL 32-1999, Section 1.2).
- References to the International Code of Marketing of Breast-milk Substitutes (1981) and the Global Strategy for Infant and Young Child Feeding and World Health Assembly resolution WHA54.2 (2001) in the Standard for Infant Formula and Formulas for Special Medical Purposes Intended for Infants (Codex Stan 72-1981 Rev.1-2007, Section 1.4), and the Standard for Processed Cereal-Based Foods for Infants and Young Children Codex (Stan 074-1981, Rev. 1-2006, Section 1). The International Code includes 'advertising' within the definition of 'marketing'.
- Reference to advertising in the Guideline for Use of Nutrition and Health Claims as cited above.

One reason for the reluctance of the Commission to enter into the field of advertising may be that in many countries advertising is regulated under legislation pertaining to general consumer protection and trade practices for goods in general rather than under food law. It may be that the question is not one of legal competence in the matter, but of technical competence and a division of responsibilities at the national level being reflected in the Commission's debates.

2.9 Conclusions

The Codex General Standard for the Labelling of Prepackaged Foods was conceived as the key Codex text for consumer protection within the sense of protecting the consumer against fraud, deception and economic disadvantage; in other words 'ensuring fair practices in the food trade'. For certain foods, those for which the

Commission has elaborated commodity standards, the General Standard is supplemented by specific labelling provisions, including the *Name of the Food* and, in these cases, the General Standard and the commodity standard must be read together. The WTO has affirmed importance of these labelling provisions in the case of EC-Sardines (WTO, 2002). An extension of this concept has led to the development of guidelines in prevent the misuse of certain claims: health and nutrition claims, claims that a food is 'organic' or 'Halal'.

From the original concept of the General Standard as primarily a trading standard, the Commission has adopted a series of amendments, revisions and additional interpretative texts that add to the information content of labelling, especially in date-marking and nutrition labelling. Thus the idea of a standard preventing fraud and deception gradually became expanded to include requirements aimed at allowing consumers to make informed choices when purchasing food.

The Commission and its Committee on Food Labelling are now grappling with the question of including measures that implement public policy on diet, physical activity and health, some aspects of which clearly fall within the Commission's established technical competence and some of which require further expansion of this competence. The history of food labelling in Codex suggests that the Commission will take positive but cautious steps to meet this challenge.

2.10 References

[Note: CAC stands for Codex Alimentarius Commission; CCFL stands for Codex Committee on Food Labelling; ALINORM is an symbol for documents within Codex]

CAC (1969). Report of the Sixth Session of the Joint FAO/WHO Codex Alimentarius Commission, ALINORM 69/67. Rome: FAO/WHO, para 127–129.

CAC (1989). Report of the Eighteenth Session of the Joint FAO/WHO Codex Alimentarius Commission (ALINORM 89/40). Rome: FAO/WHO.

CAC (2003). Recommended International Code of Practice – General Principles of Food Hygiene CAC/RCP 1-1969, Rev. 4-2003. Rome: FAO/WHO.

CAC (2008a). Guidelines for Use of Health and Nutrition Claims (CAC/GL 23-1997). Rome: FAO/WHO.

CAC (2008b). Relations between Commodity Committees and General Committees: Food Labelling. Procedural Manual of the Codex Alimentarius Commission. Rome: FAO/WHO, 104–105.

CCFL (1972). Report of the Seventh Session of the Codex Committee on Food Labelling, ALINORM 72/22. Rome: FAO/WHO.

CCFL (1979). Report of the Thirteenth Session of the Codex Committee on Food Labelling, ALINORM 79/22. Rome: FAO/WHO.

CCFL (1980). Report of the Fifteenth Session of the Codex Committee on Food Labelling, ALINORM 81/22, Appendix IV. Rome: FAO/WHO.

CCFL (1994). Report of the Twenty-Third Session of the Codex Committee on Food Labelling, ALINORM 95/22. Rome: FAO/WHO.

CCFL (2000). Report of the Twenty-Eighth Session of the Codex Committee on Food Labelling, ALINORM 01/22. Rome: FAO/WHO.

CCFL (2004). Discussion Paper on Advertising (CL 2004/54-FL). Rome: FAO/WHO.

CCFL (2008a). Report of the Thirty-Sixth Session of the Codex Committee on Food Labelling (ALINORM 09/31/22). Rome: FAO/WHO.

CCFL (2008b). Report of the Thirty-Sixth Session of the Codex Committee on Food Labelling, ALINORM 08/31/22, Appendix IV. Rome: FAO/WHO.

CCNFSDUL (2008). Report of the 30th Session of the Codex Committee on Nutrition and Foods for Special Dietary Uses, ALINORM 09/32/26. Rome: FAO/WHO.

FAO/WHO (2003). *Joint WHO/FAO Expert Consultation on Diet, Nutrition and the Prevention of Chronic Diseases* (WHO technical report series; 916). Geneva: WHO.

UNITED NATIONS (1985). Consumer Protection. General Assembly Resolution A/RES/39/248 of 16 April 1985. Geneva: UN Department of International and Social Affairs. Available at: http://www.un.org/documents/ga/res/39/a39r248.htm [accessed July 2009].

WHO (2004). Global Strategy on Diet, Physical Activity and Health – Resolution WHA57.17. Geneva: World Health Organization. Available at: http://www.who.int/dietphysicalactivity/goals/en/index.html [accessed July 2009].

WTO (2002). European Communities – Trade Description of Sardines. Geneva: World Trade Organization.

3

International legal frameworks for food labelling and consumer rights

Margret Vidar, Food and Agriculture Organization of the United Nations (FAO), Italy

Abstract: The chapter discusses the key international human rights questions of relevance to labelling, explains the consumer protection considerations behind labelling and analyses in some detail the most relevant provisions of international trade law.

Key words: food labelling, international human rights law, consumer rights, consumer protection, international trade law, international legal frameworks for food labelling and consumer rights.

3.1 Introduction

Food labelling frameworks aim to regulate different interests, which range from human health and consumers' rights to international trade. National labelling laws must therefore seek a balance of these interests and take account of different international legal obligations that may condition national frameworks.

Labelling rules can be divided into mandatory and voluntary labelling rules. The former determine which information must always be displayed on labels; the latter govern information that may be displayed. For both types of labelling there may be rules about conformity assessments.

The right to health and the right to adequate food are among the key recognized human rights that have a bearing on food labelling, along with the right to information and participation. Consumers' rights include the right to receive adequate and complete information to make their own choices and to handle the food safely. This implies a duty of the state to guarantee that the information displayed on labels is accurate, sufficient to ensure the safety of the product, enables traceability and permits tracing responsibilities.

International trade rules recognize the right of countries to preserve human, animal and plant health and to pursue legitimate goals, such as the protection of consumers against deceptive practices. At the same time, labelling requirements and certification constitute barriers to free trade, and can be particularly hard for developing countries to comply with. Therefore, such requirements should be proportionate to the objective they serve, transparent and not discriminate between countries.

Food labelling thus touches a number of branches of international law, from human rights law to environmental law to trade law and international food standards. These are sometimes presented as opposing branches of law; in particular, there is perceived conflict between human rights and trade.

However, in international law, there is a strong presumption against normative conflict (International Law Commission, 2006, paras 37–43). When negotiating new agreements that create obligations for them, states are generally assumed not to wish to create conflict with existing obligations. The International Court of Justice referred to this presumption in the *Right of Passage* case and stated, 'it is a rule of interpretation that a text emanating from a Government must, in principle, be interpreted as producing and intended to produce effects in accordance with existing law and not in violation of it' (ICJ, 1957: 142).

This principle is also valid for successive international obligations. The Vienna Convention of the Law of Treaties provides rules on application of treaties including rules on observance, retroactivity and application of successive international agreements, preventing conflicts of implementation.

It must therefore be assumed that states intend different branches of international law that have a bearing on labelling of goods to be in harmony and not in conflict with one another. This takes on a particular significance when it comes to international trade law and international human rights law. They must be presumed to be harmonious with and should be seen in the context of one another.

In the following, this chapter will discuss the international human rights of relevance to food labelling and will also highlight some of the more relevant consumer protection issues. It will then analyse in some detail the most relevant provisions of international trade law, and the objectives they seek to pursue. The conclusion will return to the question of perceived or actual conflict between the different branches of international law.

3.2 International human rights law

International human rights law is linked to labelling laws with regard to the right to food, the right to information and the right to participation, amongst others. Environmental law, including for protection of endangered species, sustainable production methods and more, is also sometimes reflected in labelling to address consumer concerns.

Human rights law provides general principles against which to judge labelling provisions and the processes leading to their adoption at the national and international levels. They do not provide detailed labelling provisions themselves and may not always solve conflicts between two or more valid principles, such as the right to adequate food, the right to information and the right to affordable food.

3.2.1 Right to food and health

Food labelling laws can be seen as one of the ways a state protects the human right of individual consumers to adequate food. The right to adequate food is recognized in article 25 of the Universal Declaration of Human Rights (UDHR) and is binding on the 160 state parties to the International Covenant on Economic, Social and Cultural Rights (ICESCR), where the right is recognized in article 11 as part of the right to an adequate standard of living and separately as the 'fundamental right to be free from hunger'.

The right implies the right to produce one's own food or to purchase affordable food that forms part of a healthy and balanced diet. State parties must respect, protect and fulfil this right progressively (CESCR, 1999, para 15). This implies that the state must first respect people's existing access to food, second protect this right from third party infringement, primarily by the enactment and enforcement of laws, and, finally, the state has an obligation to fulfil the right to adequate food by creating an enabling environment for people to feed themselves in dignity and by providing for transfers of food or cash to buy food when people are unable to do so through their own efforts. Food labelling laws are an expression of the state obligation to protect the right to adequate food.

The adequacy standard refers to the safety, nutritional value and cultural acceptability of the food (CESCR, 1999, paras 9–11). Labels on safe handling and storage of food as well as on the content and nutritional value are therefore of direct relevance and so are labels relating to cultural beliefs or traditions, for instance on whether meat is 'halal'. It might also be argued that acceptability of GMOs in food is a cultural question, and therefore should be covered by labels from a human rights point of view in those countries. At the same time, the right to food implies that food must be economically accessible (*ibid*, para 13), so affordability remains an important consideration in labelling regulations, as labelling requirements may incur costs to the producers and subsequently to the consumer.

Article 12 of the ICESCR recognizes the right to highest attainable standards of health. The current obesity epidemic is creating new pressures on regulators to take action to protect consumers from nutrient-poor and energy-dense foods, through restrictions on marketing and through new labelling schemes aimed at assisting consumers in making healthier choices. These developments gain legitimacy through the right to health as well as the adequacy element of the right to food.

The Voluntary Guidelines to Support the Progressive Realization of the Right to Adequate Food in the Context of National Food Security (Right to Food Guidelines for short), were adopted by the FAO Council in November 2004 by consensus and contain provisions on labelling principles in Guideline 9 on Food Safety and Consumer Protection:

> 9.7 States should adopt measures to protect consumers from deception and misrepresentation in the packaging, labelling, advertising and sale of food and facilitate consumers' choice by ensuring appropriate information on marketed food, and provide recourse for any harm caused by unsafe or adulterated food, including food offered by street sellers.

The Guidelines also mention labelling in the context of their nutrition guideline, stating in paragraph 10.2: 'States are encouraged to take steps, in particular through education, information and labelling regulations, to prevent over-consumption and unbalanced diets that may lead to malnutrition, obesity and degenerative diseases.'

According to the World Conference of Human Rights, 'all human rights are universal, indivisible and interdependent and interrelated' (UN, 1993, para 5). In the context of food labelling, the right to information is intimately linked to the right to adequate food. Thus, individuals have a right to accurate information about the food they buy.

3.2.2 Right to information

The right to information is recognized in UDHR (article 19) as 'the right to freedom of opinion and expression; this right includes freedom to hold opinions without interference and to seek, receive and impart information and ideas through any media and regardless of frontiers.' The 164 state parties to the International Covenant on Civil and Political Rights (ICCPR) are obliged to enforce this right, as it is recognized in similar terms in ICCPR article 19. According to the UN Special Rapporteur on the right to freedom of expression, '[a]lthough international standards establish only a general right to freedom of information, the right of access to information, especially information held by public bodies, is easily deduced from the expression 'to seek [and] receive … information' as contained in articles 19 of the [UDHR] and [ICCPR]' (UN, 2004, para 39). Many countries now have laws that stipulate the right to access public information and the obligation to disclose it (*ibid*, para. 37).

In the case of food labelling standards, however, the information is mostly not held by public authorities but by food producers. The standards are thus public law that regulates private law interaction between buyers and sellers. In human rights terms, as with the right to food, and closely related to the right to food, the state has the obligation to protect the right to information through adequate legislative provisions and other measures. In this context, freedom of expression becomes important as well, as that right can be considered to also have a negative aspect, i.e. a right not to speak (Miskiel, 2001: 227). The right of the consumers to get information may then be limited by the right of the producers not to provide information. In the balancing of these aspects, considerations relating to the right to adequate and affordable food and public health, as well as environmental protection, would be very important.

The Convention on the Rights of the Child recognizes the right of the child to the highest attainable standard of nutrition, and, in this context, contains a provision on the promotion of breastfeeding (article 24), which is scientifically recognized as the best nutrition for babies. Article 9 of the Code of Conduct on Marketing of Breast-milk Substitutes (adopted by WHO in 1981) contains strict provisions on labelling of infant formula. The label must contain a statement on the superiority of breastfeeding and 'neither the container nor the label should have pictures of infants, nor should they have other pictures or text which may idealize the use of infant formula'.

This issue highlights the links between labelling and other methods of marketing a food product, as the Code of Conduct prohibits advertising of breastmilk substitutes and limits marketing methods strictly, so as to avoid displacement of breastfeeding in cases where formula is not needed.

3.2.3 Right to participation

Human rights, such as the right to food and the right to information, also have a process element, such as accountability, non-discrimination and recourse. Participation is another important principle in the realization of human rights, and is also a human right in itself. Thus, ICCPR article 21 recognizes the right to peaceful assembly; article 22 the right to freedom of association; and article 25 the right to take part in the conduct of public affairs. The principle of participation means that people should be able to determine their own well-being and participate in the planning, design, monitoring and evaluation of decisions affecting them. Individuals must be able to take part in the conduct of public affairs, including the adoption and implementation of policies (FAO, 2009). In this regard, all individuals are consumers at some point and to a greater or lesser degree.

The principle of participation is implemented in practice through the inclusion of consumer groups and manufacturers in the shaping of policies and legislation on labelling and other such issues. In a democratic society, participation is also ensured through the elected representatives, who are then entrusted with taking decisions and adopting laws.

3.2.4 Right to environment

The right to food is limited in principle to food acquisition in ways that are sustainable (CESCR, 1999, para 8). This implies a limit to the ways in which food is produced so as not to threaten the right to food of future generations.

The 1972 United Nations Conference on the Human Environment declared that 'man's environment, the natural and the man-made, are essential to his well-being and to the enjoyment of basic human rights – even the right to life itself' (para 1). The 1992 Earth Summit adopted the Rio Declaration and Agenda 21, which contain numerous references to human rights in connection to environmental protection and sustainable development.

While a human right to sustainable environment is a relative newcomer to the human rights field, environmental law has a separate existence through international legal developments over the last few decades. Labelling is related to environmental law for instance in the case of labels such as 'dolphin-safe tuna', or living modified organisms regulated by the Cartagena Biosafety Protocol.

3.3 Consumer protection

The Latin maxim *caveat emptor* means 'let the buyer beware' and is no longer considered appropriate for retail selling of food. Rather, it is recognized that consumers often face imbalances in economic terms, educational levels and bargaining power. The consumer should be protected from unsafe food, unfair practices and inaccurate or misleading information. Measures should also be taken to promote more sustainable consumption patterns.

Many countries have adopted consumer protection laws, both general laws and for specific sectors. Consumer protection concerns are also incorporated directly into food labelling provisions, or at least form the underlying basis for the provisions, whether on health and safety, accuracy of information or mandatory disclosure. Consumer protection and rights are often influenced by human rights: they share the same underlying values, whether or not there are explicit human rights references in the relevant provisions. This is the case for the right to food, the right to health, the right to information and the right to participation.

Consumer protection is recognized to some extent as a legitimate objective of measures that may constitute barriers to trade (see subsequent section). The exact balance from a trade perspective is being established on a case by case basis in the WTO.

There is no international legally binding instrument that details consumer protection as such. The UN Guidelines for Consumer Protection (as expanded in 1999) were adopted by the General Assembly and fall within the so-called soft law category: they rely on persuasion rather than the force of law. They are the most thorough international document on the issue. The document states that consumers should have the right to access non-hazardous products as well as the right to promote just, equitable and sustainable economic and social development and environmental protection (article 1). It espouses the principles of consumer protection with regard to health and safety, and economic interests. The principles also include access to adequate information to enable consumers to make informed choices according to their individual wishes and needs, consumer education and the availability of effective consumer redress. Furthermore, the freedom to form consumer groups is recognized (article 2).

The Consumer Protection Guidelines contain a number of provisions, including for the provision of information necessary to enable consumers to take informed and independent decisions, as well as measures to ensure that the information provided is accurate (article 22). It is, however, interesting to note that they do not refer directly to human rights.

The concept of sustainable consumption is a particular focus of the UN

Guidelines since their expansion in 1999. The issue was well covered in the 1992 Rio Declaration on Development and the Environment and Agenda 21. The guidelines state that unsustainable patterns of consumption, particularly in industrialized countries, are the major cause of environmental degradation (article 4). They further state that policies for sustainable consumption should take into account the goals of eradicating poverty, satisfying basic human needs of all members of society and reducing inequality between and within countries.

Measures to promote sustainable consumption, especially those that empower consumers and demand their choices to be ethical, including through food labelling, should be distinguished from other, more drastic measures, such as banning of marketing of products that fall short of standards. For instance, it is illegal to buy and sell products from certain endangered species, rather than it being allowed and subject to a labelling requirement.

The Consumer Protection Guidelines (article 24) specifically refer to 'voluntary and transparent eco-labelling programmes' among the many provisions aimed at environmental aspects related to consumer protection.

Ecological, fair trade and similar labels measures spring from sustainable consumption concerns and are backed by international human rights and environmental law, as well as the goals set out in international conferences from Rio to Rome, for a more equitable and just society. If consumers are to be able to influence methods of food production and choose foods that have smaller ecological impact and that promote equity and social justice, then producers must be either forced or allowed to provide the information consumers need. This is done through labelling as well as advertisements and other ways of communication.

It should be noted, however, that the Consumer Protection Guidelines also contain a specific provision that consumer protection methods should be consistent with international trade obligations (article 10). This means that the Guidelines could not be used to challenge international trade law. However, human rights law and environmental law, as binding legal standards, could possibly be used for interpretation and application of international trade law, in the spirit of seeking convergence between different international obligations. The Consumer Protection Guidelines could also serve to interpret ambiguous provisions of international trade law.

3.4 International trade agreements

Labels serve many different purposes in international trade. They transmit information to the importing country and the foreign purchaser, contain essential safety and health information and support consumers' choice. They further permit tracing the origin of the products and thus maintain public order and safeguard the rights of consumers to seek redress. On the other hand, stringent labelling requirements that are different for each country hinder international trade. Trade agreements impose limits on labelling requirements countries can make, in order to facilitate international trade.

This section examines labelling from an international trade law perspective. It is important to distinguish between the following national regulatory aspects:

1. Legally binding (obligatory) labelling requirements: These determine which information it is obligatory to provide on a food label in a country.
2. Voluntary labelling schemes of public or private origin: Such systems may be regulated: (a) through framework law or regulation (ensuring legal protection and furthering the policy interests involved), with details in non-binding standards; or (b) based on private standards, under the framework of the national general labelling or consumer protection legislation.
3. Legal frameworks for conformity assessment, which regulate accreditation, monitoring and surveillance of certification bodies as well as the recognition of the equivalence of foreign conformity assessment bodies.

In the following, some of the key issues and principles of the WTO agreements will be analysed, with a view to clarifying what labelling requirements are consistent with international trade rules. Domestic legislation on labelling should be shaped with those rules in mind, so as to increase competitiveness of exporters and to avoid trade disputes with other WTO members.

3.4.1 The relevant trade agreements

The Marrakech Agreement establishing the World Trade Organization (WTO Agreement) entered into force on 1 January 1995, along with a slate of revised and new trade agreements annexed to it. As of July 2008, WTO had 153 members and many more applicants. The WTO provides a forum for negotiating agreements aimed at reducing obstacles to international trade and ensuring a level playing field for all, thus contributing to economic growth and development. Multilateral treaties are seen as fairer than bilateral trade treaties, where parties are often in very unequal positions. They facilitate international trade by creating a rules-based environment and more harmonized standards for exporters.

Food labels come under the ambit of a number of WTO agreements, notably the General Agreement on Tariffs and Trade (GATT), the Agreement on the Application of Sanitary and Phytosanitary Measures (SPS); the Agreement on Technical Barriers to Trade (TBT), and the Agreement on Rules of Origin. These agreements provide the legal ground rules for international trade applicable to labelling provisions.

GATT provides the key principles of WTO law on international trade in goods. The more specific agreements, such as TBT and SPS, prevail in case of conflict (WTO Agreement Annex 1A, General interpretative note). The TBT Agreement aims to ensure that product requirements, including labelling requirements, and procedures that are used to assess compliance with those requirements, do not create unnecessary obstacles or 'technical barriers' to trade. It covers all goods, whether agricultural or industrial.

The SPS Agreement contains specific rules for the drafting of national provisions aiming at the protection of human life, plant or animal health from diseases and health hazards, which may constitute barriers or establish restrictions to trade. Its objective is two-fold: (a) to recognize the sovereign right of members to determine the level of health protection they deem appropriate; and (b) to ensure that a sanitary or phytosanitary requirement does not represent an unnecessary, arbitrary, scientifically unjustifiable or disguised restriction on international trade.

If a measure falls under the SPS Agreement, then the TBT Agreement is not applicable (article 1.5 TBT). Sanitary measures on food labelling include indications and conformity marks to the effect that a product complies with a certain provision on microbiological criteria, pesticides residues or food additives. Health warnings, allergen information, expiry date, handling and storage information also belong to this category (see Annex A SPS). Other measures, which cannot be considered sanitary or phytosanitary measures (even those with the objective of protecting human health), such as list of ingredients and nutrients (fats, proteins, carbohydrates, etc.), fall within the scope of the TBT Agreement. In other words, the SPS Agreement covers food safety aspects, rather than health aspects related to food composition or balanced diets.

Each WTO member sets laws and regulations on food imports, including labelling requirements. They must ensure that these do not create unnecessary barrier to trade. Any food labelling measure or law that may hinder free trade has to stand up against the scrutiny of the WTO agreements.

WTO members have a legal obligation to comply with WTO provisions in the drafting and implementation of their food labelling schemes. The Dispute Settlement Body receives complaints from member about non-compliance. It consists of the entire membership, so is served by a smaller Appellate Body which in turn constitutes a Panel for each case. Labelling provisions found to be inconsistent with WTO rules could cause fines, sanctions and commercial embargoes. The possible economic consequences strengthen observance of international trade law, which can be contrasted with other fields of public international law with weaker compliance mechanisms, for instance human rights law.

3.4.2 Principles of international trade

The WTO agreements share a number of common trade principles, which are all applicable to labelling provisions. These are:

Non-discrimination

Members agree to apply technical regulations equally to domestic and imported products without any differentiation between the two, and to imported products from different member countries. These principles are aimed to ensure the fair and undistorted competition among countries in their trade relations.

Harmonization

Members agree to use relevant international standards as a basis for preparing and harmonizing technical regulations and standards.

Equivalence
Members agree to recognize technical regulations different from their own if they fulfil the same policy objectives.

Mutual recognition
Members agree to enter into negotiations with other members for the mutual recognition of conformity assessment procedures including testing, inspection, calibration and certification.

Transparency
Members agree to notify WTO organs of measures and ensure domestic transparency of procedures.

Proportionality
Members agree that technical regulations measures should not restrict trade more than necessary to achieve legitimate objectives. Members may establish mandatory labelling requirements to pursue legitimate objectives such as consumer protection, provided that such measures do not unnecessarily restrict trade.

Special and differential treatment
WTO Agreements generally recognize the particular trade, development and financial needs of developing country members, including least developed countries, and provide that they be treated differently in some respects, allowing for exceptions, flexibility and transition periods as well as technical assistance.

3.4.3 Mandatory and voluntary labelling requirements

National legislation requires certain information to appear on food labels, such as the name of the food, ingredients of processed food, the name of the producer, etc. It will then set limits and frameworks for voluntary labelling, that is, information that producers *may* provide, and under what circumstances they may make certain claims, such as health claims or other statements that serve to market the food and give consumers more information. The main principle is generally to protect consumers against misleading information while providing for consumer choice and a healthy marketing environment. There are some differences between country practice as to whether an issue is covered by mandatory or voluntary labelling, and this is sometimes a source of contention, for instance regarding the subject of genetically modified organisms. Food safety labelling information that comes under the SPS agreement is normally mandatory.

The TBT Agreement applies to compulsory labelling requirements (technical regulations) and voluntary labelling indications (standards), whether they are developed by governments or private entities, at the national or the regional level. The difference between a technical regulation and a standard is elaborated in Annex 1 to the TBT Agreement:

> 1. *Technical regulation*
> Document which lays down product characteristics or their related processes and production methods, including the applicable administrative provisions, with which compliance is mandatory. It may also include or deal exclusively with terminology, symbols, packaging, marking or labelling requirements as they apply to a product, process or production method.
> 2. *Standard*
> Document approved by a recognized body, that provides, for common and repeated use, rules, guidelines or characteristics for products or related processes and production methods, with which compliance is not mandatory. It may also include or deal exclusively with terminology, symbols, packaging, marking or labelling requirements as they apply to a product, process or production method.

Mandatory labelling provisions, such as product name, list of ingredients, weights and traceability information, are always regulated by 'technical regulations'. WTO case law has further defined what falls under technical regulation and what under the definition of a standard (WTAB/R, 2001, para 68).

Public or private rules that cover labelling in a non-binding, voluntary fashion are considered 'standards' for the purposes of TBT. These voluntary labels can be ruled by non-binding instruments or by legally binding provisions establishing the legal frameworks for producers who may want to adhere, voluntarily, to a certain system of labelling. Regulating voluntary labels by legally binding instruments has the advantage of strengthening legal security to consumers and operators, with instructions on the use of the indications defined by specific rules, rather than general legislation on consumer protection and labelling. This is the case for instance for laws on organic production that impose a general ban on marketing of foods labelled as organic unless specified conditions are met.

For both mandatory and voluntary claims, governments have the duty to monitor the claims and protect consumers' right to receive accurate and true information.

Labelling requirements falling within the definition of a *technical regulation* are subject to relevant TBT provisions, including provision of information (article 10); technical assistance (article 11) and special and differential treatment (article 12). These labelling measures must also meet the requirement of TBT article 2, which can be summarized as follows:

1. must not discriminate against imported like products;
2. must not be more trade restrictive than necessary to fulfil a legitimate objective (proportionality);
3. must be monitored and reviewed to address changes in circumstances and objectives;
4. when available and where appropriate, must adopt international standards as its basis;
5. must be notified to other members through the Secretariat; and

6. must when appropriate specify technical regulations based on product requirements in terms of performance rather than design or descriptive characteristics.

Labelling requirements falling within the definition of a 'standard' are subject to some provisions of the TBT Agreement and to the provisions of the Code of Good Practice for the Preparation, Adoption and Application of Standards, contained in Annex 3 of the TBT Agreement. A number of these provisions reflect those for technical regulations, including those on non-discrimination, necessity, the use of international standards, technical assistance and special and differential treatment. In general, the requirements for standards are less stringent than for technical regulations.

3.4.4 Relevant international standards

Both the SPS and TBT Agreements encourage the international harmonization of food standards – including for food labelling – and cite international standards, guidelines and recommendations as the preferred measures for facilitating international trade in food.

Both the TBT Agreement (article 2.4) and the SPS Agreement (article 3) provide a presumption that measures based on international standards, guidelines or recommendations are consistent with the respective agreement. WTO Members may establish their own (higher) level of health protection for labelling under the SPS Agreement, but they must be based on science-based risk assessment (article 2.2 SPS).

The SPS Agreement (preamble and article 3) explicitly recognizes as relevant international standard setting bodies Codex Alimentarius Commission, the body set up by the International Plant Protection Convention (IPPC) and the World Organisation for Animal Health (OIE). Chapter 1 focuses on Codex standards. This presumption is not the same as the standards being legally binding. The WTO Appellate Body has clarified: 'Articles 3.1 and 3.3 of the SPS Agreement permit a Member to depart from an international standard if the Member seeks a higher level of protection, the level of protection pursued is based on a proper risk assessment, and the international standard is not sufficient to achieve the level of protection pursued' (WTAB/R, 2002, para 273).

Unlike the SPS Agreement, the TBT Agreement does not identify which standard setting organizations are considered as relevant. However, WTO case law has explicitly confirmed that Codex Alimentarius standards can be relevant (WTAB/R, 2002, paras 287–291). The case law has further clarified:

1. that it is not necessary that the standard is approved by consensus (hence by the entire international community) to be considered as relevant international standard (WTAB/R, 2002, paras 222–223);
2. relevant should be understood as 'bearing upon, relate or be relevant for the purposes (of interpreting a technical regulation under question)' (WTAB PR, 2002, para 7.68 and WTAB/R, 2002, para 233).

Members can depart from relevant international standards that are 'ineffective' or 'inappropriate' means for the fulfilment of the legitimate objectives pursued through the technical regulation (article 2.4 TBT). Means are 'ineffective' if they do not have the function of accomplishing the legitimate objective pursued (based on results), and they are 'inappropriate' when they are not specially suitable for the fulfilment of the legitimate objective pursued (based on the nature), according to WTO case law (WTAB PR, 2002, para 7.116 and footnotes 91–92. See also WTAB/R, 1998, para 165).

Members may adopt SPS measures, including labelling provisions, which result in higher levels of health protection than the international standards provide – or measures aimed at health concerns for which international standards do not exist – provided that they are scientifically justified. Measures may not be more trade restrictive than necessary to achieve the 'appropriate level of protection', taking into account the technical and economic feasibility of alternative measures (article 5 SPS).

3.4.5 Legitimate objectives of labelling requirements

Given that all labelling requirements can hinder free trade, international trade law only allows national labelling requirements that serve legitimate objectives. This section explores what these may be.

Article XX of GATT (also known as *chapeau* clause) includes a list of ten permitted exceptions to the principles of free trade set forth in the Agreement. The following are those relevant to labelling:

> b) necessary to protect human, animal or plant life and health;
> d) necessary to secure compliance with laws and regulations which are not inconsistent with the provisions of this Agreement, including those relating to (...) the protection of patents, trade marks and copyrights, and the prevention of deceptive practices;
> g) relating to the conservation of exhaustible natural resources if such measures are made effective in conjunction with restrictions on domestic production or consumption.

WTO case law demonstrates that the exceptions should be interpreted in a narrow manner (*US – Shrimp*, WTAB/R, 1998, para 35). Furthermore, it has determined that 'when considering a measure under article XX, we must determine not only whether the measure *on its own* undermines the WTO multilateral trading system, but also whether *such type of measure*, if it were to be adopted by other Members, would threaten the security and predictability of the multilateral trading system' (WTAB PR, 1998, para 7.44).

The exceptions listed in GATT article XX are allowed as long as the resulting measures are not unjustified or arbitrary. This implies a condition that the country does not have different means of pursuing those goals that would avoid trade restrictive practices. In this framework, the general principles of international law, and other international agreements ratified by the members, can also be considered for interpreting the extension of an exception (*US – Shrimp*, WTAB/R, 1998, para 35. See also Vienna Convention on the Law of Treaties article 31.3.c).

Similarly to article XX of GATT, the preamble of the TBT Agreement recognizes members' rights to take the necessary measures to achieve a number of policy objectives such as 'the quality of its exports, the protection of human, animal or plant life or health, of the environment, or for the prevention of deceptive practices, at the levels it considers appropriate'.

Article 2.2 of the TBT Agreement identifies as legitimate objectives '*inter alia* national security requirements; the prevention of deceptive practices; protection of human health or safety, animal or plant life or health, or the environment'. The article further provides an open list of relevant elements for assessing the risk, including '*inter alia*, available scientific and technical information related processing technology or intended end-uses of products'.

WTO case law has confirmed that the words '*inter alia*' in article 2.2 extend the list of legitimate objectives beyond those explicitly listed. It has also determined that ensuring 'market transparency, consumer protection, and fair competition' are legitimate objectives (WTAB/R, 2002, paras 286–291).

While some areas of consumers' rights are addressed by the legitimate objective of preserving public health and preventing deceptive practices, it is not clear to which extent consumers' right to information could be considered as a legitimate objective *per se*. However, it could be argued that in certain cases consumers must be informed about processes and provenance of a product for the sake of environmental protection, which is among the legitimate objectives. It is also possible that measures that fall within the range of the UN Guidelines on Consumer Protection (as expanded in 1999) would be considered a legitimate objective, but there is as yet no case law that would confirm or deny this.

Another interesting question related to legitimate objectives is animal welfare. Considering that many consumers have ethical concerns over the treatment of the animals they eat, without this necessarily affecting animal or human health, or the environment, the question of whether consumers have a right to know about this arises.

As mentioned above, OIE standards are recognized as relevant international standards on animal health under the SPS Agreement and could presumably then also be considered relevant for the purpose of the TBT Agreement and the determination of legitimate objective under article 2.2. The 2008 version of the Terrestrial Animal Health Code of the OIE includes in Section 7 a number of standards on animal welfare. Some of these standards have direct, clear linkages with animal health, while other measures have different objectives related to ethics or to improvements of production. It is unclear (a) whether labelling requirements based on such ethical reasons would be considered legitimate objectives under article 2.2 TBT, and (b) whether the OIE Terrestrial Animal Health Code would be considered a relevant international standard in this regard. It should be noted also that little international consensus on animal ethics exists, as opinions vary according to national systems, infrastructure and traditions. In any case, most labelling regarding animal welfare is voluntary. (On the issue of animal welfare standards and WTO see Thiermann and Babcock, 2005.)

3.4.6 Non-discrimination

One of the cornerstone principles of international trade for the purposes of labelling legislation is the principle of non-discrimination (most-favoured-nation and national treatment) laid down in GATT articles I and III and reiterated in the SPS and TBT Agreements, among others. This implies that like products should be treated equally irrespective of their origin. According to the principle of national treatment, foreign products cannot be treated differently than a local product based on its origin (except for specific import measures). The most-favoured-nation principle implies that any requirements must not discriminate between like products from different countries.

With respect to labelling, non-discrimination means that members can only require those labelling standards of foreign goods that are applicable for national products. If a member applies a labelling requirement to imports from one country it has to apply equal requirements to 'like products' imported from other countries. Labelling regulations must therefore be clear and detailed enough; for instance, language requirements have to be explicitly made of national products if they are to be applied to foreign products.

The concept of 'like products' is very important for determining whether a measure is discriminatory. WTO has determined that the definition should be construed narrowly and on a case by case basis (WTAB/R, 1996: 20–23).

There are a number of relevant factors for the analysis of whether one product is like to another: (a) the product's end-uses in a given market and whether it can serve the same or similar end-use; (b) consumers' tastes and habits and the extent to which consumers perceive and treat the products as alternative means of performing particular functions in order to satisfy a particular want or demand; (c) the product's physical properties, nature and quality; and (d) the international tariff classification of the product. It has been confirmed also that health risk is a legitimate factor in considering whether products are like (*EC-Asbestos,* WTAB/R, 2001: 101–103).

In determining whether products are like in relation to labelling requirements, it is also necessary to consider indications referring to process and production methods (PPMs). Two broad categories of PPMs can be differentiated: product-based PPMs and non-product-based PPMs. Product-based PPMs refer to production methods that affect the characteristics of the product itself. This type of PPM is found more frequently in industrial process requirements, ensuring a product's quality and fitness of use. This is the case for label marks such as 'organic' where the final products fulfil some requirements such as absence of chemically synthesized pesticide residues, or restricted use of veterinary and food processing inputs. Non-product-based PPMs are situations where the results of the production method are not transmitted by the product itself. This is the case for some environmental and social labels such as 'fair trade', and the 'dolphin-safe' label on tuna cans (see box below for the WTO case and Chapter 7 on fisheries eco-labelling).

Tuna–dolphin trade disputes

Tuna fishing methods caused incidental deaths of dolphins, which started to receive public attention in the United States in the 1980s – there were consumer boycotts mounted in protest against the fishing methods. In turn, some US canning factories took up a label of 'dolphin-safe' on their tuna cans, allowing consumers to choose the more environmentally friendly option (USDA, 2000: 22–24).

The United States then set rules for fishing methods that led to the import ban on tuna that was not certified as complying with the US standards. Mexico filed an international trade complaint in 1991 (WTO, 1991). The Panel in the case considered that an embargo fell foul of GATT but that US rules on labelling as 'dolphin-safe' were consistent with GATT because they were designed to prevent deceptive practices (article XX d). The case was settled outside of GATT, so the Panel report was not formally adopted and thus does not qualify as GATT jurisprudence (WTO environmental disputes webpage).

3.4.7 Transparency

Labelling requirements and conformity assessments must be transparent, clear and published to comply with the principle of transparency laid down in article X of GATT. WTO case law has ruled that non-transparent procedures for certification and poor notification to applicant countries were contrary to the transparency principle of GATT (WTAB PR, 1998, para 183).

Transparency and dissemination of information are also key requirements in the TBT Agreement. WTO members are obliged to notify other members through the Secretariat of all mandatory labelling requirements that are not based substantially on a relevant international standard, and that may have a significant effect on the trade of other members (article 2.9). The TBT Committee has adopted a number of recommendations and decisions concerning notification procedures for drafting technical regulations and conformity assessment procedures.

3.4.8 Conformity assessments

Conformity assessment is an essential stage of the labelling process and subsequent food trade. It ensures that the labelling of food products is accurate. Countries may require that compliance of products with their technical regulations or standards is monitored by certification bodies that ensure an equivalent level of conformity to their national conformity assessment systems. The growing complexity of conformity assessment systems threatens to introduce additional burdens, which may undermine access to markets of products, especially from developing countries. The principles of non-discrimination, prevention of unnecessary barriers to trade and technical assistance to least developed countries are applicable to conformity assessment procedures, together with specific provisions included in articles 5 to 9 of the TBT Agreement.

Conformity assessment procedures should not be stricter or be applied more strictly than necessary to give the importing member adequate confidence that products conform to the applicable technical regulations or standards, taking account of the risks non-conformity would create. This implies that the procedures should (a) be completed as expeditiously as possible and in a no less favourable order for national and imported products; (b) be written and published, including fees; (c) include remedies.

Harmonization with international standards and recognition of equivalence are applicable for both public and private conformity assessment procedures. Therefore, mutual recognition agreements between regulatory bodies (i.e. government to government) and non-regulatory bodies (i.e. private sector) are increasingly important.

3.5 Conclusions

The introduction to this chapter mentioned a perceived conflict, generally speaking, between human rights law and trade law and wondered if this was the case also for labelling requirements.

Human rights laws are not unequivocal about the right of the consumer to information. Rather, the consumer's right to know should be balanced against the cost implications of ensuring that information is provided, because the human right to food is not only the right to adequate food but the right to affordable food. It is therefore not possible to argue that human rights would always be an argument for maximum disclosure. On the contrary, most human rights advocates are worried about technical barriers to trade that developing countries face, and thereby smaller farmers and food producers and processors (UN, 2009).

International trade law recognizes the right of countries to protect human, animal and plant health and the environment, as well as to protect consumers from misleading information. This is entirely consistent with human rights. WTO case law stresses the principle of proportionality and the use of international agreements. Thus, if countries were to agree on measures to ensure that dolphins do not get killed in tuna fishing, mandatory labelling and certification is likely to be upheld under WTO agreements. In the absence of such an agreement, the use of voluntary labelling schemes goes some way towards protecting dolphins while not constituting an unnecessary barrier to trade.

One of the controversies with regard to labelling today concerns food derived from GMOs. Some trade law rulings would point to the question of whether the process of bioengineering had an effect on the product and, if not, that products would have to be treated as 'like' whether or not they were the result of bioengineering. Voluntary labelling schemes are more likely than mandatory labelling requirements to be considered consistent with WTO rules. However, from a human rights perspective, if consumers in a given culture care deeply about whether their food is bioengineered or not, it can be argued that they have a right to be informed about all products that contain GMOs, on the grounds of the cultural dimension of the right to adequate food. This could entail mandatory labelling requirements in that country. The principle of participation could also challenge the recognized bases of rules under WTO agreements: It is possible that public opinion and parliament, for whatever reason, decide that GMOs must be labelled, while SPS only accepts scientific risk assessment and TBT might also not accept a labelling measure based solely on the will of the people, without technical justification under a recognized legitimate objective. Of course, the question of costs would also have to be borne in mind, so it is not clear at all whether there could be an absolute rule on this from a human rights or consumer protection perspective.

It is also not clear yet to what extent animal welfare could be the subject of mandatory labelling requirements. It will be interesting in the future, as OIE standards on animal welfare that are not directly related to animal health are expanded and applied by some countries, and whether these could be contested as outside the scope of action considered acceptable by trade agreements.

Countries wishing to play it safe from a trade law perspective could focus on implementing the standards of Codex Alimentarius and might be advised to favour voluntary labelling schemes so as not to fall foul of the international trade law principle of proportionality. However, they might also opt for different requirements, which were primarily informed by human rights and consumer protection concerns, paying more attention to the wishes of the public and democratically elected representatives than to other concerns. They might argue that human rights law supported such approaches, in addition to this being consistent with WTO law as properly interpreted. The future may tell whether such approaches are viable under international trade law as it stands.

3.6 Acknowledgement

The author is grateful for the research and drafts provided by Carmen Bullón Caro, Legal Officer, FAO Development Law Service on international trade law.

3.7 References

CESCR (1999). *General Comment 12. The right to adequate food.* UN doc. E/C.12/1999/5, 5 May 1999. Geneva: Economic and Social Council.

FAO (2009). *Guide on Legislating for the Right to Food.* Rome.

INTERNATIONAL LAW COMMISSION (2006). *Fragmentation of International Law: Difficulties arising from the Diversification and Expansion of International Law*, Report of the Study Group of the International Law Commission, Finalized by Martti Koskenniemi, A/CN.4/L.682 , 13 April 2006. Geneva.

MISKIEL F.J. (2001). Voluntary labeling of bioengineered food: cognitive dissonance in the law, science, and public policy. *California Western Law Review*, 38(1), 223–254.

THIERMANN A.B. AND BABCOCK S. (2005). Animal welfare and international trade. *Revue Scientifique et Technique de l'Office International des Epizooties*, 24(2), 747–755.

UN (1993). World Conference on Human Rights. *Vienna Declaration and Programme of Action.* UN doc. A/CONF.157/23, 12 July. Geneva: Office of the United Nations High Commissioner for Human Rights.

UN (2004). *The right to freedom of opinion and expression.* Report of the Special Rapporteur, Ambeyi Ligabo. UN doc. E/CN.4/2005/64, 17 December. Geneva: Office of the United Nations High Commissioner for Human Rights.

UN (2009). *Report of the Special Rapporteur on the right to food, Olivier de Schutter.*

Addendum. Mission to the World Trade Organization. UN doc A/HRC/10/5/Add.2, 4 February. Geneva: Office of the United Nations High Commissioner for Human Rights.

USDA (2000). *Economics of Food Labeling*. By Elise Golan, Fred Kuchler, and Lorraine Mitchell with contributions from Cathy Greene and Amber Jessup. Agricultural Economic Report No. 793. Washington, DC: Economic Research Service, United States Deoartment of Agriculture, 22–24.

WTO (1991). Environment: Disputes 4. Mexico etc versus US: 'tuna-dolphin'. Geneva: World Trade Organization. Available at: http://www.wto.org/english/tratop_e/envir_e/edis04_e.htm [accessed on July 2009].

3.7.1 International treaties and soft law instruments

AGREEMENT ON THE APPLICATION OF SANITARY AND PHYTOSANITARY MEASURES, adopted 15 April 1994, entry into force 1 January 1995. Marrakesh Agreement Establishing the World Trade Organization, Annex 1A, 1867 UNTS 493. WTO members 153 on 20 April 2009.

AGREEMENT ON TECHNICAL BARRIERS TO TRADE, adopted 15 April 1994, entry into force 1 January 1995. Marrakesh Agreement Establishing the World Trade Organization, Annex 1A, 1868 UNTS 120. WTO members 153 on 20 April 2009.

CARTAGENA PROTOCOL ON BIOSAFETY, Montreal, 29 January 2000, entry into force on 11 September 2003, 2226 UNTS 208. State parties 156 on 20 April 2009.

CODE OF CONDUCT ON THE MARKETING OF BREAST-MILK SUBSTITUTES, adopted by the World Health Assembly resolution WHA34.22 on 21 May 1981.

CONVENTION ON THE RIGHTS OF THE CHILD, adopted by the General Assembly resolution 44/25 on 20 November 1989, entry into force on 2 September 1990, 1577 UNTS 3. State parties 193 on 20 April 2009.

GENERAL AGREEMENT ON TARIFFS AND TRADE, adopted on 15 April 1994, entry into force on 1 January 1995. Marrakesh Agreement Establishing the World Trade Organization, Annex 1A, 1867 UNTS 187. WTO members 153 on 20 April 2009.

INTERNATIONAL COVENANT ON CIVIL AND POLITICAL RIGHTS, adopted by General Assembly resolution 2200A (XXI) on 16 December 1966, entry into force 23 on March 1976, 999 UNTS 171. State parties 164 on 20 April 2009.

INTERNATIONAL COVENANT ON ECONOMIC, SOCIAL AND CULTURAL RIGHTS, adopted by General Assembly resolution 2200A (XXI) on 16 December 1966, entry into force on 3 January 1976, 993 UNTS 3. State parties 160 on 20 April 2009.

MARRAKESH AGREEMENT ESTABLISHING THE WORLD TRADE ORGANIZATION, adopted 15 April 1994, entry into force on 1 January 1995. 1867 UNTS 3, WTO members 153 on 20 April 2009.

RIO DECLARATION ON ENVIRONMENT AND DEVELOPMENT, AND AGENDA 21 (1992), Report of the United Nations Conference on Environment and Development. UN doc A/CONF.151/26, 12 August.

TERRESTRIAL ANIMAL HEALTH CODE (2008), 17th edition incorporating modifications agreed during the 76th General Session of the World Organization for Animal Health (OIE) International Committee in May 2008.

UNITED NATIONS CONFERENCE ON THE HUMAN ENVIRONMENT (1972), Report. UN doc A/Conf. 48/14/Rev. 1 (1973).

UNITED NATIONS GUIDELINES FOR CONSUMER PROTECTION (as expanded in 1999), adopted by United Nations General Assembly Decision 54/499 on 22 December 1999.

UNIVERSAL DECLARATION OF HUMAN RIGHTS, adopted and proclaimed by General Assembly resolution 217 (III) 'International Bill of Human Rights' on 10 December 1948.

VIENNA CONVENTION OF THE LAW ON TREATIES, 1969, done at Vienna on 23 May 1969, entry into force on 27 January 1980, 1155 UNTS 331.

VOLUNTARY GUIDELINES TO SUPPORT THE PROGRESSIVE REALIZATION OF THE RIGHT TO ADEQUATE FOOD IN THE CONTEXT OF NATIONAL FOOD SECURITY, adopted by the 127th Session of the Council, FAO doc. Cl/127/REP.

Legal cases

ICJ (International Court of Justice) (1957). *Case concerning the Right of Passage over Indian Territory (Preliminary Objections) (Portugal v. India)*, ICJ Reports 1957.

WTAB PR (WTO Appellate Body, Panel Report) (1998). *United States – Import Prohibition of Certain Shrimp and Shrimp Products (US – Shrimp)*, WTO doc WT/DS58/R/Corr.1.

WTAB PR (WTO Appellate Body, Panel Report) (2002). *European Communities – Trade Description of Sardines (EC – Sardines)*, WTO doc WT/DS231/R/Corr.1.

WTAB/R (WTO Appellate Body Report) (1996). *Japan – Taxes on Alcoholic Beverages (Japan – Alcoholic Beverages)*, WTO docs WT/DS8/AB/R, WT/DS10/AB/R, WT/DS11/AB/R.

WTAB/R (WTO Appellate Body Report) (1998). *United States – Import Prohibition of Certain Shrimp and Shrimp Products (US – Shrimp)*, WTO doc WT/DS58/AB/R.

WTAB/R (WTO Appellate Body Report) (2001). *European Communities – Measures Affecting Asbestos and Asbestos-Containing Products (EC–Asbestos).* WTO doc WT/DS135/AB/R.

WTAB/R (WTO Appellate Body Report) (2002). *European Communities – Trade Description of Sardines (EC – Sardines)*, WTO doc WT/DS231/AB/R.

Webpages

- WTO agreements are available at:
  http://www.wto.org/english/docs_e/legal_e/legal_e.htm
- WTO dispute settlement gateway:
  http://www.wto.org/english/tratop_e/dispu_e/dispu_e.htm
- Human rights documents are available at:
  http://www.ohchr.org/

4

Government and voluntary policies on nutrition labelling: a global overview

Corinna Hawkes, Independent Consultant, Food and Nutrition Policy, France

Abstract: Many governments around the world have developed policies to encourage a standard, truthful and informative system for labelling nutrients on packaged foods, and government oversight is increasing. Food companies are now also developing alternatives, notably labels that depict nutritional information in a graphical form. This chapter reviews these policies and identifies key trends. An important conceptual shift is that nutrition labelling is no longer perceived solely as an information tool to ensure honest commerce, but as a health promotion tool and, for the global food industry, a marketing tool. While these trends are global, there remain large and significant differences between approaches to nutrition labelling around the world.

Key words: nutrition label, government policy, food industry, Codex Alimentarius, nutrient reference values.

4.1 Introduction

A 'nutrition label' is a panel on which nutritional information about a food product is displayed. It is usually found on the food item itself, but may also be found on a display device associated with the food, such as a menu or supermarket shelf. There are two broad types of nutrition label. The first, and traditional type, is the 'nutrition facts table', a boxed table that lists the nutrients found in the food and their amount. The second, and much more recent type, is the graphical nutrition label, which displays nutritional information in a more graphical, interpretative way. Many countries and, more recently, many food companies have developed regulations, standards or guidelines to define if and when nutrition labels should be applied, their category and format, the nutrients required, and on what types of foods. Governments have historically been concerned with the nutrition facts tables, with graphical labels being developed by the food industry, although some governments are also active in this area.

The basic aim of nutrition labelling is to guide the selection of food products by consumers. While food companies can apply nutrition labels at their own volition, there are several reasons why governments have found it important to develop regulations and standards on nutrition labelling, as follows:

- to provide a standard format for labelling nutrients, thus preventing the use of a potentially confusing multitude of different formats by different food companies;
- to ensure that food companies label the 'less desirable' nutrients (e.g. saturated fats) as well as 'positive' nutrients (e.g. vitamins);
- to provide proof that nutrition claims made on the label are honest and truthful;
- to ensure that nutrition labelling does not describe a product or present information about it which is in any way false, misleading or deceptive;
- to encourage food manufacturers to apply sound nutrition principles in the formulation of foods;
- to encourage the use of a format which is effective and encourages consumers to make healthier dietary choices;
- to meet the nutrition labelling requirements of other countries, thus facilitating the export of domestically-produced foods.

Food companies come from a different perspective when developing voluntary guidelines on nutrition labelling. Their aims are generally to (i) contribute to efforts to promote healthier diets; (ii) introduce a new marketing tool and a new form of competitive advantage; and/or (iii) deflect the development of mandatory government standards.

This chapter presents a global overview of regulations, standards and guidelines on nutrition labelling around the world. It first examines government regulations on nutrition facts tables. It describes the different approaches to requirements for nutrition facts tables, including considerations of format and the foods covered. The information on government regulations was obtained using the methods described in a benchmark survey conducted in 2003/04 (Hawkes, 2004), coupled with a survey of the Food and Agriculture Organization (FAO) focal points for the Codex Alimentarius conducted in 2008. Information was obtained for 79 countries.

The chapter then examines the increasing number of voluntary guidelines developed by the food industry on graphical approaches, and the role of government in these approaches. It then identifies and analyses key trends in the regulations, standards and guidelines on nutrition labelling, followed by some concluding comments.

Nutrition Facts

Serving Size 1 cup (228g)
Servings Per Container 2

Amount Per Serving	
Calories 250	Calories from Fat 110
	% Daily Value*
Total Fat 12g	18%
Saturated Fat 3g	15%
Trans Fat 1.5g	
Cholesterol 30mg	10%
Sodium 470mg	20%
Total Carbohydrate 31g	10%
Dietary Fiber 0g	0%
Sugars 5g	
Protein 5g	
Vitamin A	4%
Vitamin C	2%
Calcium	20%
Iron	4%

*Percent Daily Values are based on a 2,000 calorie diet Your Daily Values may higher ot lower depending on you calorie needs.

	Calories	2,000	2,500
Total Fat	Less than	65g	50g
Sat Fat	Less than	25g	25g
Cholesterol	Less than	300mg	300mg
Sodium	Less than	2,400mg	2,400mg
Total Carbohydrates		300g	375g
Dietary Fiber		25g	30g

Fig. 4.1 Example of a nutrition facts table: United States. Image provided by the Food and Drug Administration (www.fda.gov).

4.2 Nutrition facts tables

Found on the back or side of food packages, nutrition facts tables comprise a list of nutrients, their amounts and some form of numerical quantifier (Fig. 4.1). Government regulations around the world dictate if and when nutrition facts tables (sometimes called 'nutrition facts panels' or 'nutrition information panels') are required, the nutrients that must be listed, the reference quantifier and the foods to which they must be applied. Each is now discussed in turn.

4.2.1 General requirements

With regard to if and when nutrition facts tables are required, countries tend to fall into one of the following categories of regulation, in increasing degree of stringency:

- no regulation (i.e. nutrition facts tables are entirely voluntary and no particular nutrient list or format is required);
- guidelines on format and nutrient list for voluntarily applied nutrition labels;
- voluntary except on foods with special dietary uses (e.g. infant formula, cereal based food for young children, diabetic food; fortified or enriched foods), in which case there are requirements for the nutrient list and format;
- voluntary unless a nutrition or health claim appears on the food, in which case there are requirements for the nutrient list and format; these regulations are often in addition to the requirement to label foods with special dietary uses;
- mandatory on all packaged foods (with some variations on the food groups covered).

Table 4.1 shows the prevalence of these regulations for 79 countries. The majority of these regulations (46 countries, or 58%) require that packaged foods carry a nutrition label only when a nutrient (or health) claim is made; some of these countries also require nutrition labels on foods with special dietary uses. An equal number of countries were identified to have mandatory labelling and no regulation (13 in both cases), but this equal number is most likely due to the over-representation of countries with some form of regulation in the sample. In reality, more countries have no regulation than mandatory labelling. Eight countries require labelling just on foods with special dietary uses.

The majority tendency for countries to require nutrition labelling when a claim is made is a reflection of guidelines from the Codex Alimentarius Commission. The Guidelines on Nutrition Labelling (CAC/GL 2_1985, revised 1993) state that nutrition labels should only be required when a nutrition claim is made. The General Standard for the Labelling of and Claims for Prepackaged Foods for Special Dietary Use (Codex Stan 146_1985) also recommends that all foods for special dietary uses display a nutrition label.

It should be noted that the absence of regulation or policy does not necessarily mean absence of nutrition labels. For example, in Jamaica, there is no requirement for nutrition labelling, but labels are widely used, largely because of the need to meet export standards (CAC, 2008a). It is likewise important to note that, even where nutrition labels are only required where a claim is made or on foods with special dietary uses, the regulations usually also set out standards for the label format when they are applied on a voluntary basis. Thus even if the label is applied voluntarily, it still must follow mandatory standards on its format.

Table 4.1 lists the countries falling into each of these categories. As indicated, there are some regional similarities in regulations on nutrition facts tables. Both Canada and the United States require mandatory labelling. The United States was one the first countries in the world to make nutrition labelling mandatory (implemented in 1994 following legislation in 1990). The only change in the regulation in the United States since that time has been the

Table 4.1 Categories of approaches to regulating nutrition labelling in 79 countries

Mandatory	Voluntary unless nutrition claim (or health claim) is made	Voluntary except for foods with special dietary uses	Always voluntary, but formatting standards in case of use	No regulation
Australia	All 27 European Union countries	Bahrain	Bolivia[c]	Bahamas
New Zealand	Switzerland	Jordan		Barbados
Canada	China[e]	Kuwait		Bermuda
United States	Colombia	Oman		Belize
Argentina	Costa Rica	Qatar		Dominican Republic[b]
Brazil	Ecuador	United Arab Emirates		Haiti
Chile	Egypt	Venezuela		Honduras
Paraguay	El Salvador			Bangladesh
Uruguay	Guatemala			Pakistan
Hong Kong (SAR)	Mexico[d]			Cambodia
Malaysia (on most foods)	Brunei			Kenya
Thailand (on some foods)[a]	Indonesia			Ghana
Israel	Japan			Jamaica
	Philippines			
	Singapore			
	Thailand[a]			
	Vietnam			
	South Africa			
	Tunisia			
	Turkey			

[a]Thailand is unique in having three sets of conditions triggering the requirement for a nutrition label: foods with nutrition claims, foods which utilize food value in sale promotion and which define consumer groups in sale promotion (that is, the usefulness or function, ingredients or nutrients of product to health for use in sale promotion and sales promotions that are aimed for specific consumer groups such as: students, executives, elderly groups); plus, as of 2007, a series of snack foods (fried or baked crispy potatoes, fried or baked popcorn, rice crackers or extruded snacks, toasted bread, crackers, or biscuits, and wafers).
[b]A first draft of regulations on nutrition labelling has been completed.
[c]The standards on nutrition labelling are being revised to reflect the latest Codex versions for possible national adoption.
[d]The possibility of mandatory labelling is currently being considered.
[e]Due to be implemented 1 May 2010. By legal definition, the new guideline is voluntary but, in practice, the guideline will be mandatory for all products that carry claims.

introduction of trans fat labelling in 2006. Canada introduced its regulations around a decade later (2003, with full compliance required in 2007), and included trans fats on the nutrient list.

Under European Union (EU) law, all 27 European Union countries have the same laws on nutrition labelling. Under Council Directive 90/496/EEC (as amended by Commission Directive 2003/120/EC), nutrition labelling is voluntary unless a nutrition claim is made in the labelling, presentation or advertising of a foodstuff. The Directive also lays down a standardised format for the presentation of nutritional labelling. In place for all EU members since 1990, the accession of 10 countries to the EU in 2004 had the effect of significantly increasing the number of eastern European countries applying this approach. The Directive is, however, in process of being revised; the proposed directive would make nutrition labelling mandatory for the EU countries (discussed further below).

There is a wide variation of regulations in Latin America and the Caribbean, ranging from no regulation at all to mandatory requirements. But there are some country groupings that follow relevant economic and trade agreements. All the MERCOSUR* countries require mandatory nutrition labelling (MERCOSUL*, 2008). This situation arose after Brazil introduced a mandatory labelling law in 2001. Following the passage of the law, neighbouring MERCOSUR countries raised concerns about the label being a potential barrier to trade. The four MERCOSUR countries then negotiated the issue, leading to the development of mandatory labelling in all four countries (implemented in 2006), but with some alterations to the format and the nutrients listed in the original Brazilian law. Chile is the only other Latin American country to require nutrition labelling on all packaged foods; the law containing this requirement was passed in 2006.

There is also an effort underway to regionalise nutrition labelling in Central America. The draft 'Reglamento Centroamericano de Etiquetado Nutricional' is being developed by the Central American Customs Union (Unión Aduanera Centroamericana) with input from all the relevant countries. The draft defines the rules that must be followed when foods are labelled with nutritional information, either voluntarily by the food company, or where it is required when a nutrition claim is made or on foods with special dietary uses. Thus, the regulations would define the rules that must be followed when foods are labelled with nutritional information, but it does not impose any requirements on when and if nutrition labels are required.

All the Gulf States (Bahrain, Kuwait, Oman, Qatar, Saudi Arabia and the United Arab Emirates) follow the same standard. Nutrition labelling is regulated through the Gulf Cooperation Council Standardization Organization (GSO) standard GSO 9/2007 (replacing GSO 9:1995) on the 'Labelling of prepackaged foods'. The standard requires nutrition facts panels on foods with special dietary uses. This approach is also taken in Jordan, but not in Egypt, which instead requires nutrition labelling when a claim is made.

Australia and New Zealand follow the same law on nutrition labelling. The law, developed by the inter-governmental agency, Food Standards Australia New Zealand (FSANZ), requires mandatory labelling. Thus all packaged foods have to bear a nutrition facts table.

Government regulations in the highest proportion of South East Asian countries require nutrition facts tables on foods only when a nutrition claim is made and on food with special dietary uses. Recent activity in some countries has departed from

---

MERCOSUR and MERCOSUL are Spanish and Portuguese acronyms referring to the Common Market of South America (www.mercosur.int/)

this tendency. In 2008, Hong Kong Special Administrative Region (SAR) introduced new regulations that made nutrition labelling mandatory on all packaged foods. This was subsequent to a new law in Malaysia mandating nutrition labelling for more than 50 categories of commonly consumed foods (implemented in 2005), and a notification issued in Thailand in 2007 mandating nutrition labels on selected snack foods. In 2008, the Chinese government also introduced new (non-binding) guidelines that require nutrition labels, where used, to have a standardised format.

No regulations were identified in South Asia or Africa, with the exception of South Africa and Tunisia.

4.2.2 Nutrient list

There is considerable variation between government regulations on the requirements on the nutrients that must be included in nutrition facts tables. Nutrients fall into one of three categories:

- Nutrients that must be declared at all times.
- Nutrients that must be declared if a claim is made about a specific nutrient.
- Nutrients that can be declared on a voluntary basis.

All countries without exception require that energy, plus proteins, total fats and carbohydrates (either total or available) must be declared where a nutrition label is required. This reflects the Codex recommendation that energy plus proteins, total fats and available carbohydrates should be listed on the label. Beyond this basic requirement, there is a great deal of variation. Some countries, including Costa Rica and Egypt, only require that energy and the basic three appear on the label, plus any nutrient for which a claim is made. Other countries, like El Salvador and the Philippines' also follow Codex by requiring energy and the basic three, but also require the declaration of vitamins and minerals when they are present in significant quantities. Other countries require anything from a total of four nutrients (plus energy) to 13. For example, China and the Gulf States require four nutrients, South Africa five, Australia/New Zealand six, the MERCOSUR countries seven, and Thailand 12. These different regulations require different mixes of nutrients. In addition to the basic three, the most commonly required nutrients are dietary fibre, saturated fat and sodium. Several countries also require sugars and trans fats and, to a lesser extent, cholesterol. A small number of countries require the declaration of calcium, iron, vitamin A and/or vitamin C (e.g. Colombia, Ecuador, Thailand). Thailand is unusual in requiring vitamin B1 and B2, and Ecuador unusual in requiring potassium.

Additional nutrients may also be required when they are subject to a claim or have been added to a fortified food. In many countries, this involves just adding the claimed nutrient into the facts table if it is not already listed. In others, a claim for specific nutrient triggers the requirement for an additional group of nutrients. In the EU, when a nutrition claim refers to sugars, saturated fatty acids, dietary fibre or sodium, an additional cluster of nutrients is required – sugars, fats, saturated fatty acids, dietary fibre and sodium – alongside energy and the basic three. For other claims, just energy and the basic three are required. In the MERCOSUR countries, when a claim is made for fat or cholesterol, the addition of monounsaturated fats, polyunsaturated fats and cholesterol is required alongside the mandatory total fat, trans fats and saturated fats. A similar principle applies in Australia and New Zealand, where a 'low cholesterol' claim triggers a requirement for information on the levels of cholesterol, trans, polyunsaturated and monounsaturated fatty acids.

In addition to these nutrients, some regulations provide a specific list of additional nutrients that can be declared on an entirely voluntary basis, as is the case in Canada, the EU, Indonesia, the MERCOSUR countries and recently implemented regulations in Colombia and the Philippines. Other regulations, such as those in Japan, allow any other nutrient to be declared on a voluntary basis. Some countries mandate the order in which the nutrients should appear (e.g. Japan, the United States), while others do not (e.g. Hong Kong SAR). Regulations also usually dictate the size and positioning of the nutrition label.

4.2.3 Reference unit

Nutrition facts tables list the nutrients required with the quantity of the nutrient, usually in grams or millilitres, alongside. An additional requirement included in all regulations is the use of a reference unit, i.e. the quantity of each nutrient relative to a specific reference unit printed adjacent to the nutrient list (Fig. 4.1). A reference unit is used to make nutrient information more consumer friendly: a standardised format allows for easier comparison between food items, and can indicate how much a food portion contributes to nutrient needs. Three reference units are used:

- *Per 100 g/100 ml*: This is the measure recommended by the Codex to quantify nutrients on a nutrition label, as it allows direct comparisons between products.
- *Per serving*: This measure is intended to allow the consumer to see the specific amount of a nutrient consumed in a likely serving size. If this form is used, the number of servings in the package must also be indicated.
- *Per recommended daily amount*: This is intended to help consumers understand the relationship between the nutrient content per serving of the product and targeted intakes of particular nutrients. Countries use different terms, such as 'daily value' 'recommended daily intake/amount,' 'guideline daily amount' or 'recommended energy and nutrient intake'. The Codex guideline recommendation is to use the 'percentage nutrition reference value' which was developed specifically for international application as the reference standard for Codex guidelines.

Again, there is wide variation in the reference unit adopted by different countries. Many regulations require the use of more than one unit, and may permit others on a voluntary basis. For example:

- Some countries just require the 'per 100 g/100 ml' unit with 'per serving' as a vol-untary addition in some cases, e.g. the EU Directive, Costa Rica, the Gulf States, South Africa, Vietnam and Israel. In addition, the EU Directive (in an approach also followed by Costa Rica and South Africa) requires that vitamins and minerals must be expressed as a percentage of the recommended daily amount.
- Some countries require *either* the 100 g/100 ml unit *or* the per serving approach, e.g. Brunei, Hong Kong SAR and Japan.
- Some countries require *both* 100 g/100 ml *and* per serving, e.g. Australia, New Zealand, Chile, China, the Philippines, Singapore and Malaysia.
- Some countries require per 100 g/100 ml and per serving *if* a reference serving size is provided, e.g. Tunisia.
- Some countries require per 100 ml/100 g *or* per serving, *plus* percent recommended daily amount, e.g. Thailand.
- Some countries require percent of recommended daily amount and per serving, but not per 100 g/100 ml. e.g. Canada, the United States, Colombia, Ecuador and the MERCOSUR countries.

4.2.4 Types of food

Most national regulations on nutrition labelling cover all packaged foods (often termed 'pre-packaged foods'). Thus where regulations require nutrition facts tables only where a nutrition claim is made, it refers to all packaged foods with a nutrition claim. Or, if it is mandatory, it refers to all packaged foods. Two countries take a differing approach. Malaysia and Thailand do not require nutrition labels on all packaged foods, but on a specific list of foods. In Malaysia, nutrition labelling is mandatory on a list of over 50 commonly consumed foods, falling into the general categories of prepared cereal foods and bread, flour-based pastries, cakes and biscuits, canned meat, fish and vegetables, canned fruit and various fruit juices, salad dressings and mayonnaise and soft drinks. These foods were selected because they are 'frequently consumed and in significant amounts, and are important to the community' (Food Safety and Quality Division, 2006, p. 9). In 2007, the Thai government passed a notification requiring that "fried and baked crispy potatoes, fried and baked popcorn, rice crackers and extruded snacks, toasted bread, crackers, biscuits and filled wafers" must be accompanied by a nutrition facts table (Ministry of Health Thailand, 2007).

Regulations may also exempt specific foods from labelling. Exceptions typically comprise waters, coffee/tea, vinegars, foods in packages less than a certain size, foods sold at fundraising events, foods purchased from restaurants and other catering services, and fruits, vegetables, meat and fish to which nothing has been added. In the United States, businesses with turnovers less than a specific amount are also exempt.

Of note, in the United States, while fresh foods are exempted, nutrition labelling information is nevertheless required 'to be displayed clearly at the point of purchase … or placed in a booklet, loose leaf binder, or other appropriate format that is available at the point of purchase' (Electronic Code of Federal Regulations). Also in the United States, there is a trend towards requiring the declaration of calories on restaurant foods (Center for Science in the Public Interest, 2008). As of October 2008, over 20 cities and states were considering legislation and regulations that would require fast food and other chain restaurants to provide calories and other nutrition information on menus and menu boards. In September 2008, California became the first state to require calories to be labelled on menus and menu boards of chain restaurants. This follows from legislation in three large cities: New York City, San Francisco and Seattle. In Canada, a bill requiring that chain restaurants declare nutritional information on their products put before Parliament in 2006 failed to pass.

4.3 Graphical nutrition labelling

Graphical approaches to nutrition labelling aim to increase the ability of consumers to see, read, interpret and act upon the nutritional information provided on the package. In this more interpretative approach, a graphic format is used, usually on the front of the packet or elsewhere in the field of vision, that displays and interprets the nutrition information. Because of the emphasis on visibility, this is sometimes referred to as 'front-of-pack' labelling, though in fact graphical formats can also be found in other locations apart from the front of the food package.

Graphical formats are a relatively recent phenomenon, and remain largely in the domain of western countries. Their use has been increasing in light of evidence that nutritional facts tables are insufficiently effective (Cowburn and Stockley, 2003). Unlike nutrition facts tables, guidelines on graphical formats have been largely developed by the food industry, with the important exception of traffic light labelling.

There are four broad types of graphical nutrition labels, each of which is now discussed in more detail: traffic light labelling, guideline daily amount (GDA) labels, nutrition scoring systems and calorie labelling.

4.3.1 Traffic light labelling

Traffic light labelling was pioneered in the United Kingdom. The idea was first proposed by a medical Non-Governmental Organization (NGO), the Coronary Prevention Group, in the early 1990s (Coronary Prevention Group, 1992). The government agency responsible for food, the Food Standards Agency (FSA), took up the approach in the mid-2000s in light of research that showed that consumers found existing nutritional labelling information complex and difficult to understand (FSA, 2006). Following extensive consultation, the FSA agreed a consistent approach for 'traffic light' labelling, with four core elements (Fig. 4.2): separate information on the key nutrients: fat, saturated fat, sugar and salt; use of red, amber

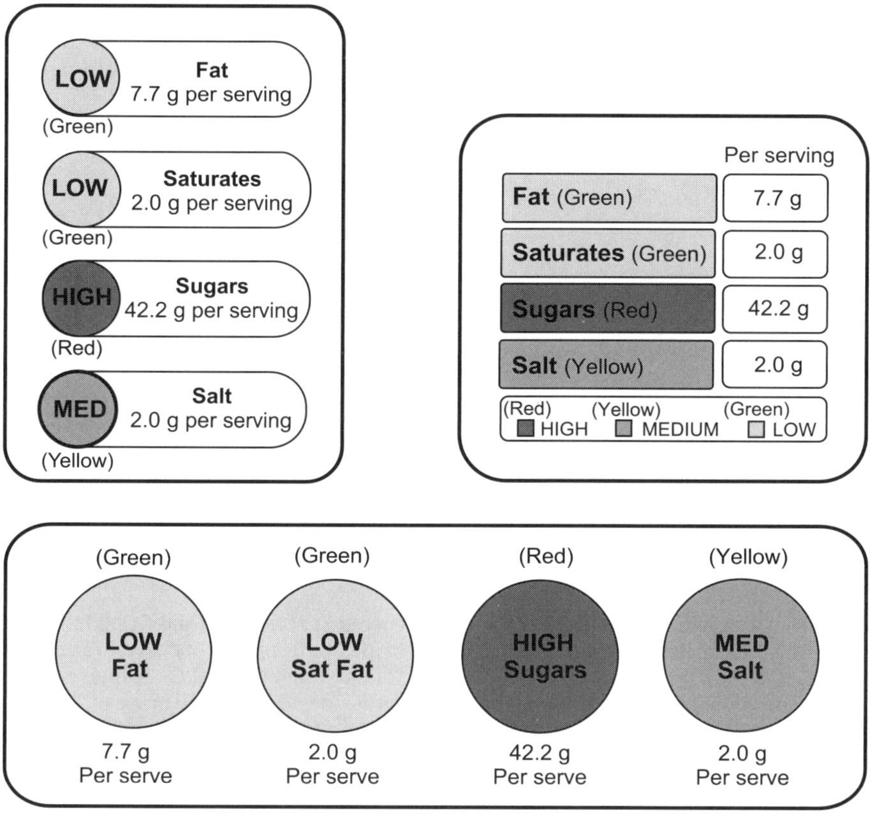

Fig. 4.2 Traffic light labels from the United Kingdom. The labels use the following colours: red – high; yellow – medium; green – low. Images provided by the Food Standards Agency (www.food.gov.uk).

or green colour coding to provide at a glance information on the level (i.e. whether high, medium or low) of the individual nutrients in the product; provision of information on the levels of nutrients present in a portion of the product; use of nutritional criteria[1] developed by the FSA to determine the colour banding.

Since the UK government does not have the authority to regulate nutrition labelling (since it falls under EU law), the FSA could not impose the scheme, but rather called on food retailers and manufacturers to adopt the approach voluntarily. The aim of the FSA was to encourage a consistent approach but with enough flexibility to allow food companies, supermarkets and restaurants to develop their own labelling schemes. They were concerned that consumers would become

[1]The nutritional criteria was based on existing advise from independent group of government advisers, the Scientific Advisory Committee on Nutrition (SACN), with the green boundaries being determined by European Regulation (EC) No 1924/2006 on Nutrition and Health claims (FSA, 2007).

(a)

(b)

Fig. 4.3 GDA nutrition labels. (a) CIAA (Confederation of the Food and Drink Industries of the EU) scheme. Reproduced with permission of the Confederation of the Food and Drink Industries of the EU (CIAA). (b) Australian Food and Grocery Council Daily Intake Guide. Reproduced with the permission of theAustralian Food and Grocery Research Council.

confused by a proliferation of schemes with differing symbols and criteria. As of April 2009, nine retailers, 31 manufacturers, five service providers and one restaurant were using labels that follow the official FSA guidance. Two retail chains in Portugal and France have also adopted the 'traffic light' approach. Governments in Chile and Thailand have also considered traffic light labelling although it was never adopted.

4.3.2 Guideline daily amount labelling

Although many UK food companies have taken up the traffic light approach, it remains unpopular with the food industry more broadly on the basis that it conveys the impression that foods are either 'good' or 'bad'. Partly as a result, the European food industry developed the Guideline Daily Amount (GDA) approach. GDA labelling involves presenting the amount of energy and key nutrients in one portion of the food as a percentage of the 'guideline daily amount' (i.e. the amount of energy/nutrients recommended that an average person consume in one day) in a graphical form, usually with some part of the label on the front of the package (Fig. 4.3). The leading food industry trade group in Europe, the Confederation of the Food and Drink Industries of the EU (CIAA), developed a GDA labelling scheme in 2006, which they advised all their members to follow. The CIAA GDA system involves labelling the calorie amount in one portion of the product and its translation into the percentage of GDA on the front of the packet, with the voluntary addition of four nutrients (fat, salt, sugar and saturated fat) either on the front or the back, depending on the packaging. The graphic takes the appearance of 'thumbnail' (Fig. 4.3a). Many main brand food manufacturers have adopted the approach. An independent survey commissioned by the CIAA in 2008 of 2026 food and drink producers in France, Italy, Spain, the United Kingdom and the Netherlands showed that nearly half of all respondents (44%) were using GDA labels (CIAA, 2008). The CIAA also estimated that, by the end of 2008, some 1030 brands, including 80% of all soft drinks and branded breakfast cereals in the EU would be using GDA labelling.

While some retailers have adopted the CIAA approach, e.g. Tesco (UK) and Aldi, Lidl and Metro (Germany), many supermarkets have designed their own GDA labels for their own-brand foods as a means of creating competitive advantage. This variation reflects advice from the trade association representing European retailers, EuroCommerce, that their members provide nutritional labels on own-brand products, but with no particular recommended format (EuroCommerce, 2007).

Governments and governmental agencies have played a limited role in the development of GDA schemes, but have sometimes been involved. The CIAA scheme and the approach take by EuroCommerce were developed as 'commitments' made to the EU Platform on Diet, Physical Activity and Health, a European Commission initiative that encourages the food industry to make measurable commitments to promoting healthy diets and addressing obesity in the EU. In addition, one country, Germany, has adopted the CIAA scheme as its official guidance to the food industry on nutrition labelling (Bundesministerin für Ernährung, Landwirtschaft und Verbraucherschutz, 2008).

A GDA scheme is also in place in Australia (Fig. 4.3b). Termed the 'Daily Intake Guide', the scheme was launched by the leading trade association representing the food industry, the Australian Food and Grocery Council (AFGC) in 2006. The label is very similar to the CIAA approach, but includes a greater number of nutrients in order to follow Australian law. Nutrition labelling is mandatory in Australia, and nutrients must be listed (in a nutrition information panel on the back of the pack) per portion and per 100 g/100 ml, with per GDA (termed percent daily intake) as a voluntary addition. If per GDA is applied, companies are permitted to apply a calories GDA label. If the company also wants to display other nutrients as GDAs, they must include the six nutrients required as mandatory on the nutrition facts panel, not just the four required by the CIAA scheme (AFGC, 2008; FSANZ, 2008). As of February 2009, the Daily Intake Guide appeared on over 1100 products produced by leading food manufacturers in Australia, such as McDonald's, Coca Cola, McCain, Birds Eye, and some food retailers, such as Woolworths and Coles.

GDA labelling is also reportedly used in other countries, including Canada, the United States and several middle income countries (e.g. see Lobstein, 2008). This is the result of initiatives taken by specific companies; there are no GDA-specific

Fig. 4.4 Example of a nutrition scoring system. Reproduced with permission of the Guiding Stars Licensing Company.

government or food industry-wide policies or schemes in place in these countries. In the United States, for example, Mars and Kellogg's apply GDA labelling, but the sugar GDA percentage value is excluded because the Food and Drug Administration has not established a GDA for sugar consumption (Kellogg, 2008; Mars, 2008).

4.3.3 Nutrition scoring

Nutrition scoring is an approach taken by retailers in the United States (Fig. 4.4). The approach was initiated by a supermarket retailer, Hannafords. Their 'Guiding Stars' scheme labels foods with either one star ('good' nutritional value) or two ('better') or three ('best') (Hannafords, 2008). The foods score is estimated using a proprietary system based on the presence of vitamins, minerals, fibre and/or whole grains and trans and/or saturated fats, cholesterol, added sugars and added sodium. No stars appear on foods that do not fall into one of the three categories. Unlike the schemes adopted by European supermarkets, it is not limited to own-brand foods, but appears on the shelf in front of any foods that qualify, including fresh foods. As of October 2008, it covered 25 000 foods.

A second 'on-the-shelf' labelling scheme is the NuVal Nutritional Scoring System (Yale Griffin Center, 2008). Unlike the Guiding Stars system, it has been developed to apply to all foods, rather than just those defined as being nutritious. Developed completely independent of industry by a research centre, NuVal scores the nutrient density of food on a scale of 1 to 100; the higher the NuVal score, the higher the nutrition value. Each food value is determined by an algorithm which analyses the composition of 30 nutrient factors in each food, including fibre, folic acid, vitamins A, C, D, E, B12, B6, potassium, calcium, zinc, omega 3 fatty acids, carotenoids, magnesium, iron, saturated fat, trans fat, sodium, sugar, cholesterol, fat quality, protein quality, energy density and glycaemic load. The intention was that NuVal scores be posted on shelf tags next to the product price, so consumers can see at-a-glance the nutritional value of the foods they buy. As of September 2008, three grocery store chains (representing over 100 stores in the Eastern and Midwestern United States) had committed to posting NuVal scores for more than a dozen food categories, including fresh protein, fresh produce, frozen vegetables, cereal, salty snacks, canned vegetables, bread, milk, cookies, crackers, eggs/egg products, drinks (shelf-stable and refrigerated), pasta and shortening/ oils.

4.3.4 Calorie labelling

GDA labelling can involve the labelling of calories on the front of food packages, but other approaches to calorie labelling are also being experimented with, mainly in the United States. For example, one state and three large cities now require chain restaurants to label calories on their menus and menu boards. In addition, some chained restaurants in the United States are adopting calorie labelling on a voluntary basis. In October 2008, Yum Brands! (the US-based company that owns Pizza Hut, KFC and Taco Bell) announced that it would place calorie information on company owned outlets throughout the United States. The company also called for federal legislation to establish uniform guidelines based on those implemented in California (Yum! Brands, 2008).

4.4 Trends

Comparison with a benchmark review published in 2004 (Hawkes, 2004) shows that there have been key changes in the policy and regulatory environment around nutrition labelling over the past few years. There have been five key trends: greater government oversight, albeit slow to develop; increased adoption of mandatory labelling; greater number of voluntary approaches by the food industry and increased use of graphical nutrition labelling, though largely limited to western countries; the application of longer *and* shorter nutrients lists; and the increased labelling of trans fats. Each of these is now discussed in turn.

4.4.1 Greater government oversight, albeit slow to develop

More governments are now deciding to regulate the provision of nutrition information on food packages. Between 2003 and 2008, at least 12 countries introduced or implemented new regulations on nutrition labelling: the four MERCOSUR countries, Chile, Colombia, China, Ecuador, Egypt, Malaysia, Thailand, Tunisia, plus many accession countries in the EU and Hong Kong SAR. Other countries, such as Bolivia and Mexico, are in the process of developing regulations. In some cases, this reflects the introduction of legislation where there was no previous guidance in this area (e.g. Colombia, Egypt); in others, the change has involved adding the requirement to declare nutritional information when a nutrition/health claim is made to existing requirements to label foods for special dietary uses (e.g. the Philippines), and in still others, the introduction of mandatory labelling on all or select packaged food (see next section).

This trend towards greater government oversight reflects the influence of two factors. The first is the Codex Alimentarius. As already discussed, the Codex Alimentarius guidelines recommend that nutrition labelling is required on foods with special dietary uses and that have a nutrition and/or health claim. The second is the development of policies to address the rising health burden created by unhealthy diets, obesity and diet-related chronic diseases. In a shift away from the narrower approach of providing labels only to provide proof of nutrition or health claims, nutrition labelling is increasingly being adopted as a policy designed to encourage healthy diets. This is leading to more countries adopting mandatory labelling and to more companies adopting graphical labelling on a voluntary basis.

4.4.2 Increased mandatory labelling

As already described, more countries now mandate nutrition facts panels on all packaged foods, or a select list of packaged foods. Since 2003, Canada, Chile, the four MERCOSUR countries, Malaysia, Thailand and Hong Kong SAR have all introduced regulations requiring mandatory labelling. This requirement has been introduced with the objective of providing greater guidance for consumers to make healthier food choices in the context of concerns about unhealthy diets and obesity. In Chile, for example, the policy was developed in the framework of their 'Global Strategy against Obesity' (EGO-CHILE) (Ministra de Salud Chile, 2006). In Hong Kong SAR, mandatory nutrition labelling was introduced on the basis that the 'provision of nutrition information on food labels is an important public health tool to promote a balanced diet' (Food and Health Bureau, 2008).

Mandatory labelling laws have not, however, been developed without controversy, and development has tended to be time-consuming and complex. In the MERCOSUR countries, negotiation about the development of mandatory labelling was protracted due to trade concerns and debate about the nutrient list (Hawkes, 2004). The mandatory labelling law introduced in Hong Kong SAR in 2008 took years to develop, and faced particular controversy about the burden it would impose on food manufacturers. In Canada, the development and implementation of mandatory labelling was reported to be 'complex, often chaotic and unpredictable' (Health Canada, 2008). A comprehensive analysis of the development of the Canadian law identified a series of barriers to the development of mandatory labelling: costs to industry (too high); proposed timelines for compliance (too short); the design of the label (considered too large); the food covered (fresh foods exempted) and nutrient list (inclusion of cholesterol) (Health Canada, 2008).

Still, the trend towards mandatory labelling looks set to continue. The proposal in the EU Directive to make nutrition labelling mandatory comes in light of research that on average just 56% of packaged foods have nutrition labels (EAS, 2004). The proposal would require the mandatory declaration of energy, fat, saturated fats and carbohydrates, with specific reference to sugars and salt (EC, 2008). If implemented, the proposed Directive would increase the number of countries with mandatory labelling by two-thirds. Echoing another trend, the proposal would also require the nutrients to be listed in the principal field of vision (i.e. with a graphical format) (EC, 2008).

The Codex Alimentarius does not currently recommend mandatory nutrition labelling, but in 2008, the Codex Committee on Food Labelling began to discuss the possibility of updating existing guidelines to recommend mandatory labelling (see CAC, 2008b, 2009).

4.4.3 Increased use of graphical nutrition labelling, but largely limited to western countries

Nutrition facts tables are focused fundamentally on the *provision* of information to consumers. The aim of this approach is to provide consumers with information to enable them to choose nutritious foods, or to verify a nutrition claim made on the label. Graphical schemes, in contrast, all aim in some way to promote and encourage the choice of 'healthier' foods, or at least to contribute to initiatives with that aim. The development of graphical labelling thus represents an important shift from the *provision* of information to the *understanding* of that information. In the short time period between 2006 and 2008, traffic lights, GDAs, nutrition scoring, calorie labelling and menu labelling have essentially redefined the nature of nutrition labelling.

Still, graphical approaches are largely limited to western countries, where diet-related problems are higher up the political and public agenda. For example, the recent announcement of calorie labelling by Yum! Brands is restricted to their outlets in the United States. This indicates that voluntary labelling initiatives are in large part driven by market and regulatory pressures. In some cases, the development of graphic voluntary approaches aims to deflect the more discriminatory 'traffic light' approach. It is notable that in Thailand, in 2006 the Ministry of Health proposed a traffic light labelling scheme for certain snack foods, but the proposal was not pursued after opposition from the United States and some other countries (USTR, 2008). Instead, the government introduced a notification requiring the application of a nutrition facts table for these foods (Table 4.1).

4.4.4 Longer and shorter nutrient lists

As requirements for nutrition facts panels have increased, greater attention has been paid to listing nutrients in addition to the basic fats, protein and carbohydrates. As already discussed, many countries now require sodium, saturated fats, trans fats (see below), dietary fibre, sugar, cholesterol and a range of vitamins. The aim of including more is to provide consumers with as much information as possible.

However, in the light of evidence that consumers find too much information difficult to interpret, graphical approaches have favoured the inclusion of fewer nutrients, and focused instead on clearer interpretation of basic nutrients. It is notable that the proposed EC Directive would require energy plus five (fat, saturates, carbohydrates with specific reference to sugars and salt) on the basis that "in selecting the mandatory elements account has been taken of research indicating that consumers can feel overwhelmed by excessive information; the scientific advice about the most important nutrients bearing a relationship to the risk of development of obesity and non-communicable diseases; while avoiding excessive burden on food businesses, in particular small and medium size enterprises" (EC, 2008).

4.4.5 More labelling of trans fats

Trans fats must now be listed on the label in an increasing number of countries: Canada, the United States, the MERCOSUR countries, Hong Kong SAR (all mandatory labelling regulations), and Colombia (when a claim is made). The increasing inclusion of trans fats comes in light of evidence that there is a link between trans fats and heart disease. Other countries have, however, decided against compulsory declaration. In the EU, the listing of trans fats on the nutrient label is not required under the current Directive unless a nutrition claim is made about them. Under the proposed new Directive, the labelling of trans fats will remain voluntary (EC, 2008). In Australia, the mandatory law only requires that trans fatty acids are declared if a claim is made for fats (see section 4.2.2) and, as of 2008, FSANZ had no plans to require the mandatory declaration of trans fatty acids. This decision was made on the basis of a review which found that the contributions of trans fatty acids to energy intakes for Australians and New Zealanders are 0.6% and 0.7%, respectively, which is well below the goal of 1% proposed by the World Health Organization (FSANZ, 2006).

4.5 Conclusions

There is huge variation in the regulations, standards and guidelines on nutrition labelling around the world. In some countries, governments have not developed any form of policy or regulation; in others, governments have developed sophisticated mandatory schemes. Likewise, in some countries, the food industry has developed a range of graphical schemes; in others, there are no such schemes. In some countries, labelling is viewed as part of a policy package to address diet-related disease and their risk factors; in others, it is viewed more narrowly as a tool for preventing deceit.

This state of affairs is at once perplexing and understandable. Perplexing because the Codex Alimentarius sets standards and guidelines for governments to follow when developing national policies. Moreover, large amounts of packaged foods are sold by global companies. Greater global consistency thus might be expected.

But this variation is also very understandable, owing to genuine differences between countries. Particularly relevant differences include:

- *The importance of packaged foods in the national diet.* Developed countries tend to consume a far higher proportion of packaged foods than developing countries (although the rate of increase of consumption is far higher in the developing world).
- *Nutritional contexts.* Differing nutrients may be lacking or excessive in national diets, and national recommended daily intakes may vary between countries.
- *Health burden.* The burden caused by unhealthy diets, obesity and other chronic disease risk factors is far higher in some countries than others, and thus likely to be more of a priority in some countries than others.
- *The amount of food exported or imported.* A country that exports a large amount of packaged foods has incentive to regulate nutrition labelling in order to meet the needs of export markets.

These differences affect the development of both government standards and voluntary industry guidelines, since the latter are strongly influenced by market and regulatory pressures. It would thus be expected that, despite the existence of Codex, governments and industry in different countries would develop differing approaches.

Still, as packaged foods become ever more widely marketed around the world with a plethora of nutrition labels, and as 'health' becomes a stronger selling point for the packaged foods industry, it is likely that the clear trend towards greater government and industry oversight of nutrition labelling will continue. The notion of what nutrition labelling is for is also shifting; it is no longer viewed simply as an information tool to ensure honest commerce, but as a health promotion tool and, for the global food industry, a marketing tool. The actions taken by the global food industry have in fact been the most notable change in the nutritional labelling environment in recent years.

As the use of and regulation of nutrition labels increases, there are two critical questions to consider: First: Is consistency preferable? Or does it not matter that there is inconsistency *between* countries? Does it matter that there are differences between government and food industry approaches *within* countries? Is this preferable in light of national differences and the need for market-driven innovation in product development? Second, as the problem of diet-related ill-health becomes ever greater worldwide, does the regulation of nutrition labelling actually help promote healthier diets? Or are consumers as confused as ever about what they should be eating for better health?

Governments – and industry – need to consider these questions as they develop and monitor policies on nutrition labelling. For into the future, it is predictable that nutrition labelling will become an inescapable part of the global food industry's way of doing business, and of government efforts to ensure that the consumer is fully informed about the foods they eat.

4.6 References

AFGC (2008). *Daily Intake Guide Labelling Scheme.* Barton ACT: Australian Food and Grocery Council. Available at: http://www.afgc.org.au/cmsDocuments/Daily%20 Intake%20Guide%202008.pdf [accessed July 2009].

BUNDESMINISTERIN FÜR ERNÄHRUNG, LANDWIRTSCHAFT UND VERBRAUCHERSCHUTZ (2008). *Leitfadern fur erweiterte Nahrwertinfomationen auf vorverpackten Lebensmitteln.* Bonn/ Berlin.

CAC (2008a). Agenda Item 6 CX/LAC 08/16/8-Add.1 October 2008. Joint FAO/WHO Food Standards Programme, FAO/WHO Coordinating Committee For Latin America and The Caribbean, 16th Session, Acapulco, Mexico, 10–14 November 2008. Nutritional Issues Within The Region, Comments In Response To Cl 2008/16-Lac, Part D From Barbados, Brazil, Haiti and Jamaica. Rome: FAO/WHO.

CAC (2008b). Codex Alimentarius Commission Thirty-first Session, Geneva, Switzerland, 30 June–4 July 2008, Report of the Thirty-sixth Session of the Codex Committee on Food Labelling, Ottawa, Canada, 28 April–2 May 2008. Joint FAO/WHO Food Standards Programme. Rome: FAO/WHO.

CAC (2009) Codex Alimentarius Commission Thirty-second Session, Rome, Italy, 29 June–4 July 2009, Report of the Thirty-seventh Session of the Codex Committee on Food Labelling, Calgary, Canada, 4–8 May 2009. Joint FAO/WHO Food Standards Programme. Rome: FAO/WHO.

CENTER FOR SCIENCE IN THE PUBLIC INTEREST (2008). Menu Labeling website. Available at: http://www.cspinet.org/menulabeling/ [accessed July 2009].

CIAA (2008). GDA nutrition labels gaining ground throughout Europe, survey shows. Press Release July 2 Brussels: Confederation of the Food and Drink Industries in the EU. Available at: http://www.ciaa.be/documents/press_releases/pr_20080702-2.pdf [accessed July 2009].

CORONARY PREVENTION GROUP (1992). *Just Read the Label: Understanding Nutrition Information in Numeric, Verbal and Graphic Format.* London: HMSO.

COWBURN G, STOCKLEY L (2003). *A systematic review of the research on consumer understanding of nutrition labelling.* Brussels: European Heart Network. Available at: http://www.ehnheart.org/files/consumer%20nutrition-143058A.pdf [accessed July 2009].

EC (2008). Proposal for a regulation of the European Parliament and of the council on the provision of food information to consumers. Brussels, 30.1.2008, COM(2008) 40 final 2008/0028 (COD). Brussels: Commission of the European Communities.

ELECTRONIC CODE OF FEDERAL REGULATIONS. Title 21: Food and Drugs, PART 101 – FOOD LABELING Subpart A – General Provisions. 101.9 Nutrition labeling of food Paragraph a(2). Available at: http://ecfr.gpoaccess.gov/cgi/t/text/text-idx?c=ecfr&sid=563 f0b6235da3f4c7912a64cbceec305&rgn=div8&view=text&node=21:2.0.1.1.2.1.1.6 &idno=21 [accessed July 2009].

EUROCOMMERCE (2007). EuroCommerce recommendation to its members to develop Nutritional Labelling on its own-brand products. Brussels. Statement available at: http://www.eurocommerce.be/content.aspx?PageId=40733 [accessed July 2009].

EAS (2004). The introduction of mandatory nutrition labelling in the European Union. Impact assessment undertaken for DG Sanco, European commission. Brussels: European Advisory Services Available at: http://ec.europa.eu/food/food/labellingnutrition/nutritionlabel/ impact_assessment.pdf [accessed July 2009].

FISCHER H (2008). State program labels healthy menu items. *Arizona Daily Star* March 25.

FOOD AND HEALTH BUREAU (2008). Legislative Council Brief. Public Health And Municipal

Services Ordinance (Cap. 132) Food And Drugs (Composition And Labelling) (Amendment: Requirements For Nutrition Labelling and Nutrition Claim) Regulation Hong Kong: Government of the Hong Kong SAR. Available at: http://www.fhb.gov.hk/download/press_and_publications/otherinfo/080215_labelling/e_legco_brief.pdf [accessed July 2009].

FOOD SAFETY AND QUALITY DIVISION (2006). *Guide to Nutrition Labelling and Claims (as of April 2006).* Putrajaya: Ministry of Health Malaysia.

FSA (2006). Parliamentary Briefing March 2006: Front Of Pack Food Labelling Making Healthier Choices the Easy Choices. London: Food Standards Agency. Available at: http://www.epolitix.com/Resources/epolitix/Forum%20Microsites/Food%20Standards%20AgencyFinal%20briefing%20March%2006%20Signposting%20Headed.pdf [accessed July 2009].

FSA (2007). Front-of-Pack Traffic Light Signpost Labelling. Technical Guidance. Issue 2, November. London: Food Standards Agency. Available at: http://www.food.gov.uk/multimedia/pdfs/frontofpackguidance2.pdf [accessed July 2009].

FSANZ (2006). Review Report: Trans Fatty Acids. In: *The New Zealand and Australian Food Supply*. Canberra: Food Standards Australia New Zealand. Available at: http://www.foodstandards.gov.au/_srcfiles/Transfat%20report_CLEARED.pdf [accessed July 2009].

FSANZ (2008). Standard 1.2.8. Nutrition Information Requirements. In: *Australia New Zealand Food Standards Code*. ANSTAT, 2008. Canberra: Food Standards Australia New Zealand. Available at: Available at: http://www.foodstandards.gov.au/thecode/foodstandardscode.cfm [accessed July 2009].

HANNAFORDS (2008). Guiding Stars Basics. Portland, ME. Available at: http://www.hannaford.com/Contents/Healthy_Living/Guiding_Stars/faqs.shtml [accessed July 2009].

HAWKES C (2004). *Nutrition labels and health claims: the global regulatory environment.* Geneva: WHO.

HEALTH CANADA (2008). *CARMEN Pilot Canadian Case Study Final Report 'Stakeholder Convergence On Nutrition Labelling: Building Consensus On A Complex Issue'*. Government of Canada. Available at: http://www.phac-aspc.gc.ca/publicat/2009/carmen_report/pdf/CARMEN_report_final-eng.pdf [accessed July 2009].

KELLOGG (2008). *Kellogg Cereal GDAs*. Available at: http://www.kelloggnutrition.com/learn-about-labels/kellogg-gdas.html#chart [accessed July 2009].

LOBSTEIN T (2008). *Cereal Offences: A wake-up call on the marketing of unhealthy food to children*. London: Consumers International.

MARS (2008). *Mars Nutrition: What are GDAs?* Available at: http://www.marshealthyliving.com/nutritional_info/gda.html [accessed July 2009].

MERCOSUL (2008). MERCOSUL/GMC/RES No. 46/03. Regulamento Técnico Mercosul Sobre Rotulagem Nutricional De Alimentos Embalados. Monterndies: MERCOSUL. Available at: http://www.mercosur.int/msweb/portal%20intermediario/Normas/normas_web/Resoluciones/PT/Res_046_003_RTM%20Rotul%20Nutric%20Alim%20Embal_Ata%2004_03.PDF [accessed July 2009].

MINISTRA DE SALUD, CHILE (2006). Entró en vigencia nuevo etiquetado nutricional de los alimentos. Noticias 06-11-2006. Santiago: Gobierno de Chile. Available at: http://www.ego-chile.cl/paginas/not_03.htm [accessed July 2009].

MINISTRY OF HEALTH THAILAND (2007). Notification of Ministry of Public Health (No. 305) B.E.2550 (2007) Re. Labeling of some kinds of ready-to-eat foods. Bangkok. Available at: http://www.qmaker.com/fda/new/images/cms/top_upload/1224646040_Notification 305.pdf [accessed July 2009].

USTR (2008). 008 National Trade Estimate Report on Foreign Trade Barriers: Thailand. Washington DC: United States Trade Representative. Available at: http://www.ustr.gov/about-us/press-office/reports-and-publications/archives/2008/2008-national-trade-estmate-report-fo-0 [accessed July 2009].

YALE GRIFFIN PREVENTION RESEARCH CENTER (2008). NuVal™ Nutritional Scoring System Fact Sheet. Derby, CT.

YUM ! BRANDS (2008). Yum ! Brands announced US divisions will place calories on all company restaurant boards. Press Release October 1. Louisville, KY. Available at: http://www.franchising.com/pressreleases/8997/ [accessed July 2009].

5

Labelling of allergenic foods of concern in Europe

Sue Hattersley and Chun-Han Chan, Food Standards Agency, UK

Abstract: People with food allergies need to know what is in the food that they buy, in order to make safe and informed food choices. The legislation covering requirements for declaring specified allergenic ingredients used in food sold pre-packed within the European Union is described, together with best practice guidance produced by the UK Food Standards Agency (FSA) covering the provision of allergen information for foods sold non-prepacked. Further guidance from the FSA covering the management of food allergen cross-contamination, as well as new legislation covering the composition and labelling of foods for people with gluten intolerance is also explained.

Key words: allergen legislation, labelling information, guidance, Europe.

5.1 Introduction

People with food allergies need to have clear information about the ingredients used in the foods they buy so that they can successfully avoid the foods that they know they are sensitive to. Whilst avoidance of single foods that are allergenic is relatively straightforward, the increasingly complicated nature of the food supply chain means that the food allergic consumer is faced with an ever more difficult task when trying to choose foods that are safe to eat, both when buying pre-packed foods at a retail level or when eating out. For example, for an egg allergic person, avoiding eating whole eggs may be straightforward, but effectively avoiding egg used as a ingredient in a complex food (such as a glaze on top of a fruit pie) can be more challenging.

Food allergic people can react to very small amounts of the allergen they are sensitive to, sometimes to amounts as low as a few milligrams. Furthermore, the symptoms seen when an allergic reaction is triggered can range from relatively mild symptoms, such as rash, through to severe life-threatening symptoms such as swelling in the throat, difficulties breathing, collapse and anaphylactic shock. It is therefore critical that such people have accurate information about the use of allergic ingredients in a food, however low the level of use and about possible cross-contamination events during production.

There are a very large number of foods (up to 200) that have been reported to trigger allergic reactions in people around the world, but a much smaller number of foods are associated with the majority of reactions reported. The Codex Alimentarius Standard originally listed eight allergenic foods (or groups of allergenic foods) that were considered to be of the greatest public health concern (Codex Alimentarius Commission, 1985). These were:

- cereals containing gluten;
- crustaceans;
- eggs and egg products;
- fish and fish products;
- peanuts, soybeans and products of these;
- milk and milk products;
- tree nuts and nut products;
- sulphites in concentrations of 10 mg/kg or more.

It is not surprising that the most common allergenic foods will vary in different countries given the different dietary patterns. For example, the allergenic foods that have to be declared in Japan (Ministry of Health, Labour & Welfare, 2005) include buckwheat, as well as eggs, milk, peanuts and wheat (but not barley, rye or oats), but buckwheat is not included in the specified lists in the European Union (EU) or USA (EC, 2003, 2007; USFDA, 2004). However, other factors may also have an impact, for example, allergies to many fruits and vegetables are linked to pollen allergy. This type of allergy is due to similarities between the proteins present, as with the case of allergy to apple being strongly linked to birch pollen allergy, and therefore the pattern of allergies will vary depending on the flora in different countries (Vieths *et al.*, 2002; Fernández-Rivas *et al.*, 2008).

5.2 Drivers behind the development of specific EU allergen labelling legislation

In EU Member States, there has long been a requirement for labelling of ingredients used in pre-packed foods. However, the general food labelling Directive (2000/13/EC) (EC, 2000) contained a number of exemptions which meant that some ingredients were not required to be labelled.

One main exemption related to the components of compound ingredients, which themselves made up less than 25% of the final food, which did not need to be separately identified. Whilst there may be an argument that the general population did not need to have information on all the individual components of that compound ingredient, clearly for the allergic consumer it is very important to know whether wheat is present in a sausage used in a casserole or celery is present is a vegetable stock used to make a soup.

It is also important for the allergic consumer that ingredients are clearly described, so that they can determine whether the flour used to thicken a sauce is wheat flour, which would pose a risk for the wheat allergic or gluten intolerant consumer, or maize (corn) flour, which would not pose a risk, or whether the oil used in a salad dressing was made from walnut oil rather than olive oil.

It was recognised therefore, that there were situations under the general ingredients labelling legislative requirements in Directive 2000/13/EC, where the allergic consumer would not necessarily receive sufficient information. The European Commission and the Member States agreed that this deficit should be addressed by developing specific requirements that would require the clear declaration of the use of allergenic ingredients in pre-packed foods in all circumstances. This requirement for the clear declaration of allergenic food ingredients was provided by Directive 2003/89/EC (EC, 2003) which came into effect in November 2005. This legislation amended the parent Directive 2000/13/EC governing general food labelling, and therefore covers only the deliberate use of the ingredient and relates only to foods sold pre-packed.

It was agreed that the allergenic foods identified by the Codex Committee should be taken as the basis for discussion between European Union (EU) Member States and the European Commission, when the need for specific legislation on the declaration of the use of allergenic ingredients started to be considered. On the basis of scientific evidence that identified that a further three allergenic foods (sesame seeds, mustard and celery) were a public health concern in at least some EU Member States, these were added to the list of allergens to be covered by specific EU labelling requirements (EC, 2003, EFSA 2004). Subsequently, a further two allergenic foods (molluscs and lupin), were added to the EU list by Directive 2006/142/EC (EC, 2006a) on the grounds that there was evidence that these were a public health concern in EU Member States (EFSA, 2005, 2006; Radcliffe *et al.*, 2005; EC, 2006b). The current EU list of specified allergenic foods is:

- cereals containing gluten (wheat, barley, rye, oats, spelt, kamut, or their hybridised strains);
- crustaceans;
- eggs;
- fish;
- peanuts;
- soybeans;
- milk (including lactose);
- nuts (almond, hazelnut, walnut, cashew, pecan nut, Brazil nut, pistachio nut, macadamia nut and Queensland nut);

- celery;
- mustard;
- sesame seeds;
- sulphur dioxide and sulphites at concentrations of more than 10 mg/kg or 10 mg/litre, expressed as SO_2;
- lupin;
- molluscs.

It is possible that further allergenic foods could be added to the EU list in the future if there was sufficient scientific justification. An emerging new risk may arise from changes in dietary patterns, for example the introduction of a new food into the diet (such as kiwi fruit) (Lucas *et al.*, 2003, 2004; Lucas and Atkinson, 2008) or, potentially, to changes in the way food ingredients are used or processed before consumption, for example, if an extract of an allergenic food were to be used for technological purposes in a compound food where its use was unexpected.

5.3 Exemptions for certain processed ingredients derived from the specified allergenic foods

During EU negotiations on Directive 2003/89/EC, it was recognised that some ingredients derived from the specified allergenic foods would, in practice, not present an allergenic risk, due to the significant processing they undergo. It was considered that it would not be helpful for allergic consumers if such ingredients were subject to allergen labelling requirements, as this would unnecessarily restrict their food choices. In addition, it might mislead allergic consumers who inadvertently eat such products into believing that their allergy was resolving. It was therefore agreed that industry should be able to submit scientific dossiers of information to support the exemption of certain ingredients derived from the specified allergenic foods from the allergen labelling requirements. The dossiers submitted were evaluated by the European Food Safety Authority (EFSA) Panel on Dietetic Products, Nutrition and Allergies, and the Panel's opinions can be seen on the EFSA website: http://www.efsa.europa.eu/EFSA/ScientificPanels/NDA/efsa_locale-1178620753812_Opinions465.htm.

A number of dossiers were submitted and subsequently evaluated by EFSA before the deadline of November 2005 for the coming into force of the allergen labelling requirements, and some exemptions were agreed on a temporary basis, pending the submission of further supporting dossiers. These exemptions, which were set out in Directive 2005/26/EC (EC, 2005), were for a two year period. Subsequently, following evaluation of the further dossiers by EFSA, a number of permanent exemptions were set out in Directive 2007/68/EC (EC, 2007). This Directive sets out the list of all allergenic ingredients that must be declared on labels and exemptions to those declarations – see Table 5.1. This Directive came into force in November 2007 but it included a transition period lasting until 31 May 2009 to allow the food industry time to change their labelling to comply with the new provisions.

Table 5.1 Schedule of all allergenic ingredients that must be declared on labels and exemptions to those declarations (from Directive 2007/68/EC)

Allergenic ingredient	Exemptions
Cereals containing gluten (wheat, barley, rye, oats, spelt, kamut or their hybridised strains)	• Wheat-based glucose syrups including dextrose • Wheat-based maltodextrins • Glucose syrups based on barley • Cereals used for making distillates or ethyl alcohol of agricultural origin for spirit drinks and other alcoholic beverages
Crustaceans	None
Eggs	None
Fish	• Fish gelatine used as a carrier for vitamin or carotenoid preparations • Fish gelatine or isinglass used as a fining agent in beer and wine
Peanuts	None
Soybeans	• Fully refined soybean oil and fat • Natural mixed tocopherols (E306), natural D-alpha tocopherol, natural D-alpha tocopherol acetate, natural D-alpha tocopherol succinate from soybean sources • Vegetable oils derived phytosterols and phytosterol esters from soybean sources • Plant stanol ester produced from vegetable oil sterols from soybean sources
Milk (including lactose)	• Whey used for making distillates or ethyl alcohol of agricultural origin for spirit drinks and other alcoholic beverages • lactitol
Nuts (almonds, hazelnuts, walnuts, cashews, pecan nuts, Brazil nuts, pistachio nuts, macadamia nuts and Queensland nuts)	• Nuts used for making distillates or ethyl alcohol of agricultural origin for spirit drinks and other alcoholic beverages
Celery	None
Mustard	None
Sesame seeds	None
Sulphur dioxide and sulphites at concentrations of more than 10 mg/kg or 10 mg/litre, expressed as SO_2	
Lupin	None
Molluscs	None

Subsequently Directive 2007/68/EC was amended by Commission Regulation 415/2009 to extend the transition period for changing the labelling for any allergenic ingredients derived from egg and milk used as fining agents in wines until 31 December 2010, to coordinate with other changes that needed to be made to labelling of wines under Council Regulation EC 479/2008 (EC, 2008a, 2009a).

5.4 Other allergen information that manufacturers can choose to put on food packaging

5.4.1 Allergy boxes or statements

In addition to the statutory requirements to label the use of allergenic ingredients in pre-packed foods, many food manufacturers in European countries also voluntarily provide additional information on food packaging to help food allergic consumers to make safe and informed food choices. This can take the form of 'allergy advice' boxes or statements that highlight the allergenic ingredients used in the product, using phrases such as 'contains egg, milk and peanuts'. Such statements may also declare possible allergenic cross-contaminants (see Section 5.4.2). Whilst the use of such allergy statements or boxes can be used as a shortcut by allergic consumers, they are not controlled by legislation (although clearly they should not be misleading) and they should therefore not be relied upon in isolation from the ingredients list.

In addition, the use of two pieces of information on the packaging relating to allergenic ingredients does increase the chances of errors and inconsistencies in the labelling. In the UK, there have been a number of food incidents where foods have had to be withdrawn from the market as the allergens listed in an allergy statement did not match those declared in the ingredients list. In particular, it is very important that food manufacturers who choose to include such statements on a product, ensure that all the allergenic ingredients declared in the ingredients list are included in the allergen statement, as it is accepted that, despite advice to the contrary, many allergic consumers will just use such as statement as their primary source of information.

5.4.2 Allergen cross-contamination warnings

Allergen labelling legislation in the EU covers only the deliberate use of an ingredient in a pre-packed product. However, for the allergic consumer, there may be a health risk if a food product contains a significant level of an allergen as a result of accidental cross-contamination at some point in the food chain. Whilst food manufacturers can put in place a number of checks and processes to try to control the risk of the accidental presence of an allergenic food ingredient in a product, it is not always possible to completely avoid such a risk, particularly in premises that make a wide range of products, with multiple ingredients.

In such situations, many food manufacturers will opt to use some form of advisory labelling to alert allergic consumers to such risks, using phrases such '*may contain nuts*', '*made in a factory that also uses nut ingredients*' or '*not suitable for someone with a nut allergy*'. Whilst the intention of such warnings is to help allergic consumers to make safe food choices, over recent years the use of such warnings has become widespread and, for certain food products (such as biscuits, breakfast cereals and confectionery), it can be difficult to find products without such warnings (FSA, 2002). Currently there is no internationally agreed action level for cross-contamination with allergenic foods below which advisory labelling is not appropriate. Therefore manufacturers may choose to label *any* risk of allergen cross-contamination, however low or remote. In addition, improvements in allergen analytical detection methodologies have also meant that the presence of lower and lower levels of allergen can now be detected, which may also be a factor in the increasing use of allergen advisory labelling.

There is evidence from consumer research (FSA, 2005) that many food allergic consumers consider such allergen advisory warnings to be overused and therefore they are often ignored. In addition, the variety of phrases used by different food businesses can also confuse food allergic consumers who may interpret the different phrases as meaning different levels of risk. There is evidence (Hefle *et al.*, 2007) that demonstrates that 'May Contain' statements seem to be a more effective deterrent than 'shared facility' statements, with 'shared equipment' statements having an intermediate effectiveness. However, products with 'shared facility' statements were more likely to have detectable levels of cross-contamination. Food businesses are also placed in a difficult position as the ability of analytical methods to detect the presence of an unwanted food allergen continues to improve in sensitivity. The results of such tests need to be assessed in terms of what they mean in relation to risk to the allergic consumer. At present, there is little quantitative guidance available to food businesses on the management of food allergens or to inform their decision-making regarding the need for allergen advisory warnings for individual food products.

5.4.3 Development of best practice guidance on allergen management and advisory labelling issued by the UK Food Standards Agency in 2006

In the UK, there was general agreement between food allergic consumer support organisations and food businesses that the excessive use of allergen cross-contamination advisory warnings devalued the impact of such warnings and also unnecessarily restricted the choices available to food allergic consumers. The Food Standards Agency (FSA) was approached with a view to producing a single guidance document bringing together existing best practice advice on allergen management. The FSA worked with stakeholders including food manufacturers and retailers, as well as allergic consumer support organisations and food law enforcement bodies, to develop its best practice guidance on allergen management and advisory labelling guidance, which was published and was made freely available on-line in July 2006 (http://www.food.gov.uk/safereating/allergyintol/guide/). The aim of the guidance was to set out consolidated best practice advice on allergen management that would lead to the adoption of a risk-based approach for the use of allergen warning labels, thereby maintaining food safety and also helping to maximise consumer choice. The main guidance document was accompanied by a leaflet aimed at small and micro-businesses (http://www.food.gov.uk/multimedia/pdfs/publication/allergyjamjar0109.pdf) that set out the key allergy-related issues to be considered when labelling food.

At the time the guidance was produced, there was no consensus on the levels of allergenic ingredients present in foods that were likely to provoke allergic reactions in consumers sensitive to those foods. Whilst the availability of commercial test kits for detecting the presence of a number of common food allergens continues to improve, there are as yet few independently validated allergen detection methods available to food businesses and enforcement bodies. This further complicates the situation for a food business trying to control allergen cross-contamination and make decisions on whether or not advisory warnings are appropriate.

The approach taken in the guidance was to set out general principles that can be used to manage allergen cross-contamination and includes a decision tree approach to inform decision-making on whether or not advisory labelling is appropriate (see Fig. 5.1). Such decisions should be based on an analysis of the risks of unintentional allergen cross-contamination across the supply chain from agricultural production of raw ingredients through to final food product that is sold to consumers. This risk analysis comprises an assessment of the nature of the risk, whether that risk can be managed, and how the risk should be communicated, as well as involving a review process.

The nature of the risk posed in a particular situation will depend on a number of factors, including:

- the amount of the allergenic ingredient that could be present;
- the allergenicity of the particular ingredient involved (for example, refined nut oils will pose a lower risk than pieces of whole nut);
- the physical nature of the ingredients being used;
- the geography of the manufacturing environment.

Fine powders that may become airborne may represent a greater cross-contamination risk than liquid or solid ingredients although other factors, such as the ability to clean down between production runs, may be important where shared equipment is used. The risks posed by cross-contamination that is at a low level and is homogeneous throughout a product run will be different to the risks of occasional cross-contamination with discrete particles such as pieces of nut or whole seeds.

Furthermore, the risks of cross-contamination may be high at the beginning of a batch when switching between products but be insignificant later in the batch run. The risk assessment will also need to take into account the marketing of the product, such that cross-contamination will pose a greater risk in products making claims to be 'free from' particular allergens, than in general food products.

If the business determines that there is a probable risk of allergen cross-contamination that cannot be eliminated or reduced, then that risk should be communicated to the consumer via advisory labelling. It is important that such advice is clearly communicated and that it is situated close to the ingredients list on the packaging, although there should be a clear distinction between the labelling

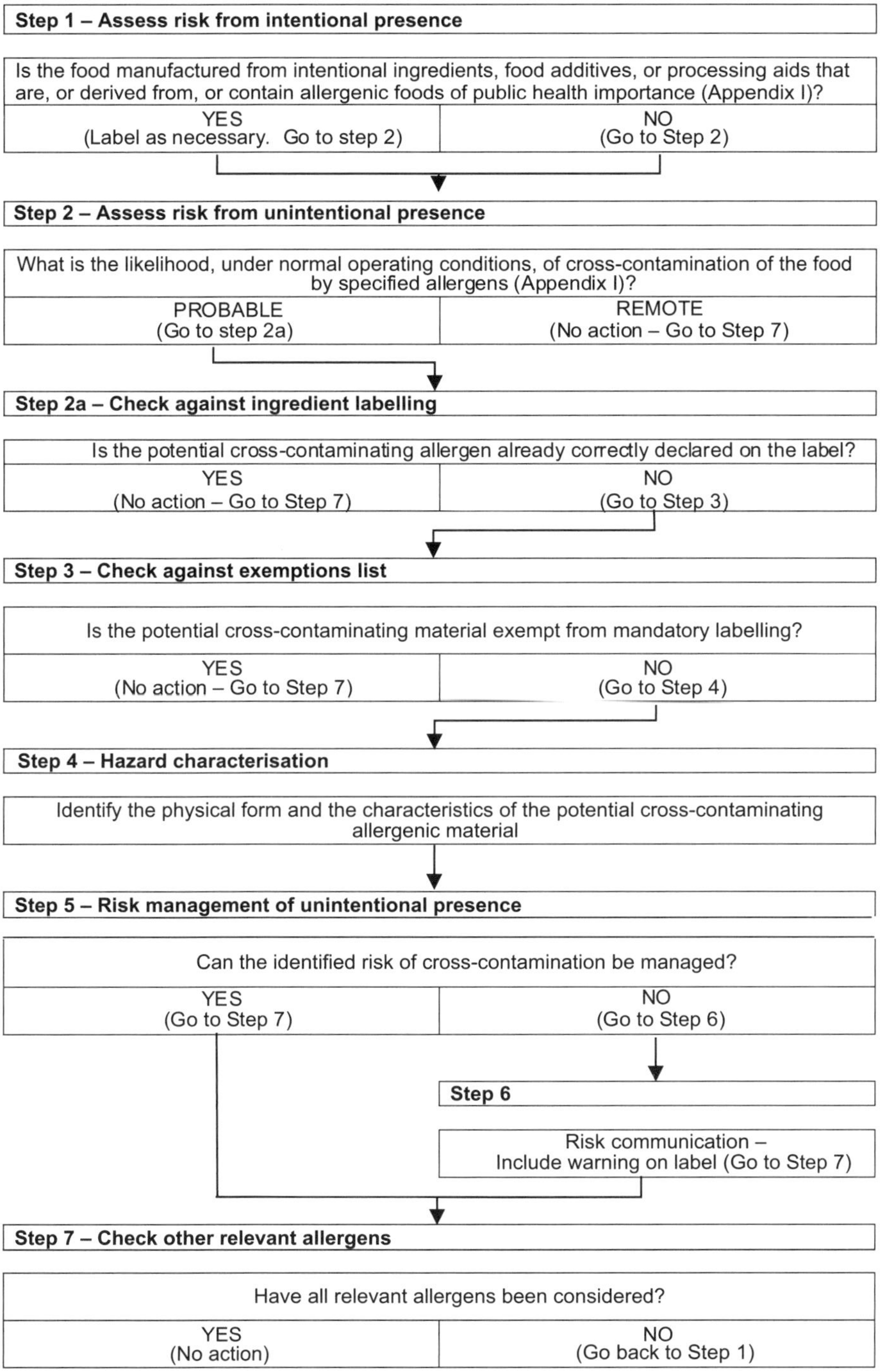

Fig. 5.1 Allergen advisory labelling decision tree.

information provided about the deliberate inclusion of allergenic ingredients that are intentionally present in the food, and those which may be there unintentionally.

In the absence of internationally agreed allergen management levels in foods to use as a basis for decisions on whether or not advisory warning labels are appropriate, the guidance advises businesses to assess whether the risk of cross-contamination is probable or remote.

Work is currently underway in a number of fora to take forward the process of setting allergen management thresholds. A workshop in Madrid in 2007 discussed with international stakeholders whether it was possible to apply risk assessment methodologies used for chemical and toxicological risk assessment to the assessment of allergenic risks. Such approaches were considered to be useful, although a number of information gaps were identified that would need to be addressed before such approaches could be used (Madsen *et al.*, 2009).

5.5 Possible legislative developments in the future, including foods sold non-prepacked

5.5.1 EU Review of food labelling legislation

In January 2008, the EU published a proposal for a Food Information Regulation (EC, 2008b) that would bring together and update a range of existing legislative requirements into a single piece of legislation. This proposal is currently being negotiated by EU Member States, as well as being considered by the European Parliament. It is anticipated that existing allergen labelling requirements will be maintained in the new Regulation, but the possibility of an extension to the current requirements to include a new requirement to provide allergen information for foods that are sold non-prepacked, including in catering situations, is also being discussed. Such an extension in the requirements is justifiable as there is evidence to suggest (Pumphrey and Gowland, 2007) that food allergic consumers are more likely to have an allergic reaction when eating out than when eating food sold pre-packed. However, due regard must be given to the ability of businesses providing food that is not pre-packed to supply such information accurately and in a way that does not impose undue administrative burdens (see Section 5.5.2).

5.5.2 Development of best practice guidance on the provision of allergen information for foods sold non-prepacked issued by the UK Food Standards Agency in January 2008

In general, food labelling legislation exempts foods sold non-packaged from the requirement to provide full ingredients listings (for example, Directive 2000/13/EC (EC, 2000). However for the food allergic consumer, there is still a need for information about the ingredients in a food to enable a safe food choice to be made. As mentioned in Section 5.5.1 above, there is evidence (Pumphrey and Gowland, 2007) that, where a person with a known food allergy who is actively trying to avoid the foods to which they react does have a further allergic reaction, this event is more commonly reported with foods that are sold unpackaged, including in catering establishments. There is a need to raise awareness in businesses selling food that is unpackaged about the needs of food allergic consumers and to provide advice to help them meet the needs of their customers.

Consumers with food allergies or food intolerances need to have information about the ingredients used in the foods they wish to purchase so that they can make safe food choices. Whilst many countries have legislation that requires the provision on such information on foods sold pre-packed, this generally does not cover foods that are sold unpackaged, including in catering establishments. This exemption arises from a consideration of the practical constraints faced by businesses selling unpackaged foods, many of whom are small or micro-businesses, as well as an acknowledgement that there is an opportunity in such transactions for the buyer to ask the person producing the food about the ingredients used.

The FSA considered that there was need for guidance for businesses selling food that is not pre-packaged, both retail and in the food service sector, to help them meet the needs of their food allergic consumers. This guidance was produced in collaboration with stakeholders from the retail sector, catering businesses and chefs, catering suppliers, food allergic consumers and enforcement bodies. It sets out a number of key messages, as well as describing examples of issues that can arise in different types of businesses providing unpackaged foods and ways that these can be addressed. The main guidance document (http://www.food.gov.uk/multimedia/pdfs/loosefoodsguidance.pdf), which is aimed at larger catering businesses and food law enforcement bodies, is accompanied by a leaflet (http://www.food.gov.uk/multimedia/pdfs/publication/loosefoodsleaflet.pdf) aimed at small and micro-businesses, as well as a poster (http://www.food.gov.uk/multimedia/pdfs/publication/thinkallergy.pdf) that can be used to facilitate staff training.

The guidance sets out three key messages for food businesses providing food that is not pre-packaged, relating to:

- effective communication, both between the customer and the business and also between the different functions within the business;
- training for staff;
- ensuring accurate information is available about the ingredients being used.

It is accepted that the allergic consumer has a responsibility when eating food that is not pre-packaged, to ask for information about the ingredients used in the product in question. Businesses should have in place a recognised procedure for dealing with such requests and staff should all be aware of that procedure. This may involve referring queries to a senior member of staff but, if information is not available, staff should never guess, and in such situations should inform the customer that they are unable to provide the information requested. It is also important that there is effective communication within a food business between the people preparing the food and those serving customers. The final decision whether or not to eat the food rests with the consumer, based on the information they receive. If it is not possible to provide allergen information about the standard products being provided, it may be possible for the business to provide alternative products not containing the allergen, such as meat cooked without the marinade or sauce or salads served without the garnishes or dressings.

All staff in businesses selling food that is not pre-packaged need to be trained to deal with requests from food allergic consumers. The type of training will vary for different types of business, but the poster developed by the FSA can be used as part of this process. Businesses also need to ensure that information about the ingredients they use can be accessed by staff and that this is kept up-to-date. Again the ways in which this is achieved will vary according the type and size of business involved. The FSA leaflet sets out seven key tips for businesses selling food that is not pre-packaged to help them help their food allergic customers.

- When someone asks you if a food contains a particular ingredient, always check every time – never guess. If you check but you're still not sure, tell the customer so they can decide for themselves.
- If you are selling a food that contains one or more of the ingredients which can cause a problem, list them on the card, label or menu – and make sure the information is accurate.
- Keep up-to-date ingredients information for any ready-made foods that you use (for example, a filling you put in a sandwich). The ingredients might be on the label or invoice.
- When you are making food, make sure you know what is in all the ingredients you use, including cooking oils, dressings, toppings, sauces and garnishes.
- If you change the ingredients of a food, make sure you update your ingredients information and tell other staff about the change.
- If someone asks you to make some food for them that does not contain a particular ingredient, don't say yes unless you can make sure that absolutely none of that ingredient will be in the food.
- If you're making food for someone with an allergy, make sure work surfaces and equipment have been thoroughly cleaned. And wash your hands thoroughly before preparing that food.

5.6 Foods sold as 'free from'

Almost all legislation governing the labelling of allergenic ingredients in foods relates to the labelling of ingredients that are deliberately incorporated into the food product and does not address claims that such an ingredient is not present. In the EU, there is the provision under the framework Directive on Foods for Particular Nutritional Uses (Council Directive 89/398/EEC) (EC, 1989) to make claims regarding the absence of gluten for foods for people with gluten intolerance (coeliac disease). For a number of years, food manufacturers within the EU, and elsewhere, could make use of the Codex Alimentarius Standard (Codex Alimentarius Commission, 1981) for foods for people with gluten intolerance, that advised that foods could make the claim that they were 'gluten free' if such foods did not contain more than 200 ppm gluten. However, scientific evidence became available (Collin *et al.*, 2004, 2007; Gibert *et al.*, 2006; Catassi *et al.*, 2007) suggesting that the 200 ppm limit for gluten did not provide sufficient protection for all coeliac patients and a revised Codex Standard was published in 2008 (Codex Alimentarius Commission, 2008).

This revised Codex Standard was the basis of discussions between the European Commission and EU Member States that resulted in the publication of Regulation EC/41/2009 (EC, 2009b) in January 2009, setting compositional standards and labelling requirements for foods for people with gluten intolerance. This regulation requires that:

- foods which are specially prepared and/or processed to meet the special dietary needs of people intolerant to gluten can make the claim 'gluten free' as long as they do not contain more than 20 ppm gluten in the food as sold to the final consumer;
- foods for people with gluten intolerance that consist of, or contain, ingredients made from gluten containing cereals (such as wheat, barley or rye) that have been especially processed to reduce gluten, can be described as 'very low gluten' provided that the level of gluten in the food as sold to the final consumer does not exceed 100 ppm;
- foods for normal consumption can be described as 'gluten free' provided that the gluten content does not exceed 20 ppm in the food as sold to the final consumer.

However, there is no other legislation that sets out requirements for foods that make claims that they are free from other allergenic foods, other than the general food law provisions in Regulation No. 178/2002 (European Communities, 2002) that, *inter alia*, prohibits unsafe food being placed on the market and that requires that food labelling should not be misleading. If food businesses want to make such claims, they need to put in place procedures and checks to ensure that they are justified.

5.7 Conclusions

In many countries, including those within the European Union, there are statutory provisions that require allergenic foods in pre-packed food to be clearly labelled where they are used as deliberate ingredients. However, at present there are no provisions within the regulations on the management of the presence of allergenic foods as a result of cross-contamination. Nor are there currently any provisions covering allergenic ingredients used in foods sold non-prepacked.

The new regulatory threshold for gluten that will control the composition of foods to be labelled 'gluten free' is a step in providing adequate protection for food allergic consumers. However, the possibility of defining regulatory thresholds for other allergens needs to be explored further. It is clear that in order to provide better food choices where 'may contain' labelled foods are concerned, the basis for making decisions on the declaration of the presence of allergens needs to be defined through establishing an allergenic management threshold. Developing workable and widely accepted allergen management thresholds requires further evidence to be used together with known clinical allergen thresholds to account for circumstances which can affect the threshold or severity of an allergic reaction. Until this is completed, it would be hard to establish any evidence-based management thresholds for controlling allergen contamination in food.

In the interim, the greatest risk to allergic consumers comes from the non-prepacked foods where currently the provision of information about the use of allergenic ingredients is not currently a legal requirement. The best practice guidance documents produced by the Food Standards Agency help food businesses to improve the management and communication of food allergens present in the foods that they sell.

5.8 References

CATASSI C., FABIANI E., IACONO G., D'AGATE C., FRANCAVILLA R., BIAGI F., VOLTA U., ACCOMANDO S., PICARELLI A., DE VITIS I., PIANELLI G., GESUITA R., CARLE F., MANDOLESI A., BEARZI I. AND FASANO A. (2007), A prospective, double-blind, placebo-controlled trial to establish a safe gluten threshold for patients with celiac disease. *The American Journal of Clinical Nutrition*, 85, 160–166.

CODEX ALIMENTARIUS COMMISSION (1981), Codex Alimentarius Commission Standard for gluten-free foods CODEX STAN 118-1981. Rome: FAO/WHO. Available at: http://www.codexalimentarius.net/web/standard_list.do [accessed July 2009].

CODEX ALIMENTARIUS COMMISSION (1985), Codex Standard 1-1985. General Standard for the labelling of prepackaged foods. Rome: FAO/WHO. Available at: http://www.codexalimentarius.net/download/standards/32/CXS_001e.pdf [accessed July 2009].

CODEX ALIMENTARIUS COMMISSION (2008), Codex standard for foods for special dietary use for persons intolerant to gluten. CODEX STAN 118-1979. Rome: FAO/WHO. Available at: www.codexalimentarius.net/download/standards/291/cxs_118e.pdf [accessed July 2009].

COLLIN P., THORELL L., KAUKINEN K. AND MÄKI M. (2004), The safe threshold for gluten contamination in gluten-free products. Can trace amounts be accepted in the treatment of coeliac disease? *Alimentary Pharmacology & Therapeutics*, 19, 1277–1283.

COLLIN P., MÄKI M.AND KAUKINEN K. (2007), Safe gluten threshold for patients with celiac disease: some patients are more tolerant than others. *The American Journal of Clinical Nutrition*, 86(1), 260.

EC (EUROPEAN COMMISSION) (1989), Council Directive 89/398/EEC of 3 May 1989 on the approximation of the laws of the Member States relating to foodstuffs intended for particular nutritional uses. *Official Journal of the European Union*, L186, 27–32. Available from: http://ec.europa.eu/food/food/labellingnutrition/nutritional/d89-398-ec.pdf [accessed July 2009]

EC (EUROPEAN COMMISSION) (2000), Directive 2000/13/EC of the European Parliament and of the Council of 20 March 2000 on the approximation of the laws of the Member States relating to the labelling, presentation and adverting of foodstuffs. *Official Journal of the European Communities*, L109, 29–42.

EC (EUROPEAN COMMISSION) (2003), Directive 2003/89/EC of the European Parliament and of the Council of 10 November 2003 amending Directive 2000/13/EC as regards indication of the ingredients present in foodstuffs. *Official Journal of the European Union*, L308, 15–18.

EC (EUROPEAN COMMISSION) (2005), Commission Directive 2005/26/EC of 21 March 2005 establishing a list of food ingredients or substances provisionally excluded from Annex IIIa of Directive 2000/13/EC of the European Parliament and of the Council. *Official Journal of the European Union*, L35, 33–34.

EC (EUROPEAN COMMISSION) (2006a), Commission Directive 2006/142/EC of 22 December 2006 amending Annex IIIa of Directive 2000/13/EC of the European Parliament and of the Council listing the ingredients which must under all circumstances appear on the labelling of foodstuffs. *Official Journal of the European Union*, L368, 110–111.

EC (EUROPEAN COMMISSION) (2006b), Standing Committee on the Food Chain and Animal Health. Section on General Food Law. Summary record of Meeting of 21 April 2006. Available at: http://ec.europa.eu/food/committees/regulatory/scfcah/general_food/summary19_en.pdf [accessed July 2009].

EC (EUROPEAN COMMISSION) (2007), Commission Directive 2007/68/EC of 27 November 2007 amending Annex IIIa to Directive 2000/13/EC of the European Parliament and of the Council as regards certain food ingredients. *Official Journal of the European Union*, L310, 11–14

EC (EUROPEAN COMMISSION) (2008a) Council Regulation (EC) No 479/2008 of 29 April 2008 on the common organisation of the market in wine, amending Regulations (EC) No 1493/1999, (EC) No 1782/2003, (EC) No 1290/2005, (EC) No 3/2008 and repealing Regulations (EEC) No 2392/86 and (EC) No 1493/1999. *Official Journal of the European Union*, L148, 1–61.

EC (EUROPEAN COMMISSION) (2008b), Proposal for a Regulation of the European Parliament and of the Council on the provision of food information to consumers. Available at: http://ec.europa.eu/food/food/labellingnutrition/foodlabelling/publications/proposal_regulation _ep_council.pdf [accessed July 2009].

EC (EUROPEAN COMMISSION) (2009a), Commission Regulation (EC) No. 415/2009 of 20 May 2009 amending Directive 2007/68/EC amending Annex IIIa to Directive 2000/13/EC of the European Parliament of the Council as regards certain food ingredients. *Official Journal of the European Union*, L125, 52–53.

EC (EUROPEAN COMMISSION) (2009b), Commission Regulation (EC) No. 41/2009 of 20 January 2009 concerning the composition and labelling of foodstuffs suitable for people intolerant to gluten. *Official Journal of the European Union*, L16, 3–5.

EC (EUROPEAN COMMISSION) (2002), Regulation (EC) No. 178/2002 of the European Parliament and of the Council of 28 January 2002 laying down the general principles and requirements of food law, establishing the European Food Safety Authority and laying down procedures in matters of food safety. *Official Journal of the European Communities*, L31, 1–24.

EFSA (EUROPEAN FOOD SAFETY AUTHORITY) (2004), Opinion of the Scientific Panel on Dietetic Products, Nutrition and Allergies on a request from the Commission relating to the evaluation of allergenic foods for labelling purposes. *The EFSA Journal*, 32, pp1–197.

EFSA (EUROPEAN FOOD SAFETY AUTHORITY) (2005), Opinion of the Scientific Panel on Dietetic Products, Nutrition and Allergies on a request from the Commission related to the evaluation of lupin for labelling purposes (Request No. EFSA-Q-2005-086) (adopted 6 November 2005). *The EFSA Journal*, 302, 1–11.

EFSA (EUROPEAN FOOD SAFETY AUTHORITY) (2006), Opinion of the Scientific Panel on Dietetic Products, Nutrition, and Allergies on a request from the Commission related to the evaluation of molluscs for labelling purposes (Request No. EFSA-Q-2005-84) (adopted 15 February 2006). *The EFSA Journal*, 327, 1–25.

FERNÁNDEZ-RIVAS M., BENITO C., GONZÁLEZ-MANCEBO E. AND DE DURANA D. A. (2008), Allergies to fruits and vegetables. *Pediatric Allergy and Immunology*, 19(8), 675–681.

FSA (FOOD STANDARDS AGENCY) (2002), *'May Contain' Labelling – The Consumer's Perspective*. Crown Copyright 2002.

FSA (FOOD STANDARDS AGENCY) (2005), *Qualitative Research into the Information Needs of Teenagers with Food Allergy and Intolerance*. Crown Copyright 2005.

GIBERT A., ESPADALER M., ANGEL CANELA M., SÁNCHEZ A., VAQUÉ C. AND RAFECAS M. (2006), Consumption of gluten-free products: should the threshold value for trace amounts of gluten be at 20, 100 or 200ppm? *European Journal of Gastroenterology & Hepatology*, 18 11), 1187–1195.

HEFLE S.L., FURLONG T.J., NIEMANN L., LEMON-MULE H., SICHERER S. AND TAYLOR S.L. (2007), Consumer attitudes and risks associated with packaged foods having advisory labeling regarding the presence of peanuts. *Journal of Allergy and Clinical Immunology*, 120(1), 171–176.

LUCAS J.S.A., LEWIS S.A. AND HOURIHANE J.O'B. (2003), Kiwi fruit allergy: A review. *Pediatric Allergy and Immunology*, 14, 420–428.

LUCAS J.S.A., GRIMSHAW K.E.C., COLLINS K., WARNER J.O. AND HOURIHANE J.O'B. (2004), Kiwi fruit is a significant allergen and is associated with differing patterns of reactivity in children and adults. *Clinical & Experimental Allergy*, 34, 1115–1121.

LUCAS J.S.A. AND ATKINSON R.G. (2008), What is a food allergy? *Clinical & Experimental Allergy*, 38(7), 1095–1099.

MADSEN C.B., HATTERSLEY S., BUCK J.,GENDEL S.M., HOUBEN G.F., HOURIHANE J.O., MACKIE A., MILLS E.N., NØRHEDE P., TAYLOR S.L. AND CREVEL R.W. (2009), Approaches to risk assessment in food allergy: report from a workshop "developing a framework for assessing the risk from allergenic foods". *Food and Chemical Toxicology*, 47(2), 480–489.

MINISTRY OF HEALTH, LABOUR AND WELFARE (2005), Food Sanitation Law. Law No. 233, December 24, 1947, last amendment Law No. 87, July 26 2005. Tokyo, Japan. Available at: http://www.mhlw.go.jp/english/topics/qa/allergies/al.html [accessed July 2009].

PUMPHREY R.S. AND GOWLAND M.H. (2007), Further fatal reactions to foods in the United Kingdom 1999–2006. *The Journal of Allergy and Clinical Immunology*, 119(4), 1018–1019.

RADCLIFFE M., SCADDING G. AND BROWN H.M. (2005), Lupin Flour Anaphylaxis. *The Lancet*, 365, 1360.

USFDA (U.S. FOOD AND DRUG ADMINISTRATION) (2004), Food Labeling and Consumer Protection Act of 2004 (Title II of Public Law 108–282). Silver Springs, MD. Available at: http://www.fda.gov/Food/LabelingNutrition/FoodAllergens/Guidancecompliance RegulatoryInformation/ucm106187.htm [accessed July 2009].

VIETHS S., SCHEURER S. AND BALLMER-WEBER B. (2002), Current understanding of cross-reactivity of food allergens and pollen. *Annals of the New York Academy of Sciences*, 964, 47–68.

6

Organic food labels: history and latest trends

Antonio Compagnoni, Institute for Ethical and Environmental Certification (ICEA), Italy

Abstract: Labelling of foods produced through organic agriculture is illustrated in this chapter. The meaning of organic agriculture is explained, with brief background information. Recent data are provided on the steady growth and diffusion of organic agriculture as an important sustainable agriculture practice and its relevance in the market place. Information is given on some of the main labelling standards, international and national, private and public, including Codex Alimentarius, IFOAM Basic Standards, EU Regulation, US National Organic Program (NOP) and Japanese JAS. The purpose and work of the FAO-IFOAM-UNCTAD International Task Force for organic regulations harmonization and equivalence is explained. Some of the main logos used in the labelling of organic products are shown.

Key words: organic agriculture, organic standard, labelling, control and certification, organic logo.

6.1 Introduction

Before describing the labelling of organic products, we should begin with an explanation of the meaning of organic agriculture, giving brief background information and recent data on its steady growth and diffusion as an important sustainable agriculture practice and its relevance in the market place. Following this brief history, some of the main labelling standards, international and national, private and public will be discussed. These include the Codex Alimentarius, International Federation of Organic Agriculture Movements (IFOAM) Basic

Standards, EU Regulation, US National Organic Program (NOP) and Japanese Organic Standard (JAS). The purpose and work of the FAO-IFOAM-UNCTAD promoted International Task Force (ITF) for organic regulations harmonization and equivalence is then briefly explained. Finally, some of the main logos for organic products are presented.

6.2 Organic agriculture definition

After a two year consultative process, the General Assembly of IFOAM adopted the Principle of Organic Agriculture in September 2005 in Adelaide, Australia. This identifies the fundamentals of organic agriculture: health, ecology, care and fairness. The General Assembly also passed a motion to establish a succinct Definition of Organic Agriculture. After almost three years of intensive work, a dedicated Task Force came up with a definition that was ratified at the General Assembly of IFOAM in June 2008 in Vignola, Italy:

> Organic agriculture is a production system that sustains the health of soils, ecosystems and people. It relies on ecological processes, biodiversity and cycles adapted to local conditions, rather than the use of inputs with adverse effects. Organic agriculture combines tradition, innovation and science to benefit the shared environment and promote fair relationships and a good quality of life for all involved. (IFOAM, 2008)

6.3 From alternative movement to international and national legislation

Organic agriculture had illustrious precursors such as the philosopher Rudolf Steiner (1861–1925), who lived in Donji Kraljevec, Dornach, father of the biodynamic movement, and other agronomists and practitioners who opposed from the very beginning the coming of age of 'industrialized chemical agriculture'. The biodynamic agriculture seal 'Demeter' appeared for the first time as early as 1927 with the first biodynamic coffee plantation at Finca Irlanda in Chiapas, Mexico. However, organic farming is a rather young phenomenon, starting in many countries in different time periods, but with very similar patterns. When agriculture became more dependent on synthetic chemicals such as fertilizers and pesticides, and more evidence was disclosed on their precise dangers to human health and to the environment, some pioneers started practising and further developing organic farming. Many of them were young people, often coming out of the agriculture world. With education, they were able to recognize the danger of chemical agriculture, pursuing organic farming as a concrete alternative. They were soon stimulated by a growing number of consumers who were ready to buy organic products and to pay premium prices to organic producers.

This started to happen in Europe and North America more or less in the same time, around the end of the 1960s and early 1970s. In 1972, IFOAM was established by five organic and biodynamic associations from France, Sweden, South Africa, the United States of America (USA) and the United Kingdom (UK). In other regions of the developed world, organic agriculture started to appear as awareness of farmers and consumers rose and they reacted to the impact of chemical industrialized agriculture. In some other regions, especially in Southern Europe, Oceania and in the developing world, another factor was, and still is, a powerful driving force: this is the northern countries' growing demand for organic foods, including the imported ones. At this stage, the labelling requirements for organic food was only a private sector issue, often with self declarations from the producers following private rules defined by a national or regional associations of organic farmers.

IFOAM in 1980 published the first Basic Standard for Organic Production and Processing, as a guiding tool for all its member organizations involved in standard setting and certification. Later on in the same decade, legislators in a few countries and regions started issuing specific norms on organic agriculture (California, France, Austria, some Italian regions). During the 1980s, organic products started to diffuse in the market place, especially in Germany, the UK and other northern European countries. The organic movement became more organized and was strengthened by a close relationship with consumers and the environmental movement. In these years the European Commission started working on a directive for organic farming (1987) that became in 1991 a Council Regulation (EC Reg. 2092/91) coming into force on 1 January 1993 (EEC, 1991). In July 1992, the Codex Alimentarius Commission decided that its Codex Committee on Food Labelling should discuss and develop the 'Guidelines for the Production, Processing, Marketing and Labelling of Organically Produced Foods' (Codex Alimentarius Commission, 1999). Soon after in Europe, organic agriculture boomed, affecting other regions as well, with a steadily growing demand for organic products in the market place. This was a result of the clear official regulatory framework, accompanied by a set of effective incentives for organic farming, through the European agro-environmental measures (EC Reg. 2078/92) (EC, 1992). In 1990 in the USA, a federal norm for organic agriculture started to develop, later becoming the United States Department of Agriculture National Organic Program (USDA, 2000). In 1999 in Japan, a specific norm was established, the JAS of Organic Agricultural Products (Japan MAFF, 2000). In the last decade many more countries followed these examples; in many cases, the main purpose was to assist their growing organic producers with obtaining access to the regulated organic markets of Europe, North America and Japan. Some large countries, such as India, China and Brazil, developed their organic norms at first for better accessing export markets, but they are now also engaged in regulating their growing organic internal markets.

In 2007–8, other international organic rules were developed, as a result of a joint effort from local governments, United Nations agencies and the private sector, due to concern about the numerous national and private standards that could eventually become a technical barrier to trade within the regions, placing unneeded restrictions on regional collaboration. Based on the international references of Codex and IFOAM organic norms, an extensive, inclusive and transparent consultation process was carried out and resulted in two regionally adapted organic norms: the East African Organic Product Standard (EAOPS) and the South Pacific Islands Countries, Pacific Organic Standard (POS) (SPC, 2008).

6.4 From niche to mainstream market

According to February 2009 data, based on FIBL (Research Institute for Organic Agriculture, Switzerland) and IFOAM research, 32.2 million hectares of agricultural land are managed organically in 141 countries (see Fig. 6.1), an increase of 1.5 million hectares compared to the previous year. In the 'World of Organic Agriculture' (Willer and Kilcher, 2009), there are over 1.2 million organic producers listed. Total turnover for organic products (food and beverages) in the world, according to a recent estimate, reached 46 billion US dollars in 2007; a three-fold increase in value from eight years before (*Organic Monitor* UK, February 2009). Most of the market share is concentrated in Europe, North America and Japan, while other regions such as Asia, Latin America, Oceania and Africa are relevant producers and exporters. However some signals of growing demand for organic products are coming from these regions too, especially in countries such as Brazil, India and China. A market niche of specialized retailers and committed customers has grown into a genuine mainstream one. Although the major retailers have larger shares of the organic market, the specialized organic stores have evolved into a dynamic and innovative sector, using franchising and marketing tools to provide a demanding public with organic food of the most diverse nature and provenance.

Another recent organic market development is the growing phenomenon of organic meals consumed outside the home. Some early experiences of organic

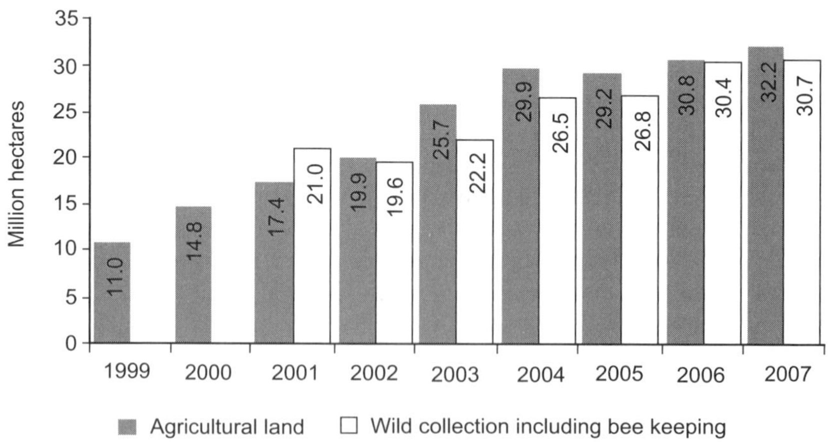

Fig. 6.1 Organic agricultural land and wild collection 1999–2007.

food in school meals had already started in Italy in the mid- and late 1980s; more recently, organic public catering is becoming increasingly relevant in many European countries in schools and hospitals, as well as private restaurants. In cases such as Italy and Spain, it is a national and/or regional law that mandates and/or supports local authorities in including organic ingredients in the menus of public schools and hospitals. Increasing and interesting developments can also be seen in the creation and consolidation of local markets for organic products all over the world, including many developing countries, especially in Latin America. Local organic farmers are becoming directly involved in large numbers of dedicated markets, special delivering schemes, and community supported agriculture experiences where citizens are actively involved together with local organic farmers. Some of these experiences are newfound ways of providing confidence and trust in the organic products trade, building concrete innovative verification schemes that are identified as 'participatory guarantee systems'.

6.5 Main normative frame

6.5.1 Common principles and general requirements

All regulations prescribe that any production and placement of labelled organic products on the market must comply with a certification process. Conventional farmers must first undergo a conversion period, with minimums ranging from one to three years before they can begin producing agricultural goods that can be marketed as organic. This period depends on the crops: if they are annuals, the period is shorter; if perennials, the period is longer. If they wish to produce both conventional and organic produce, they must clearly separate these two operations throughout every stage of production. Both farmers and processors must at all times respect the relevant rules contained in the regulations. They are subject to inspections by approved certification bodies or authorities to ensure their compliance with organic legislation. After the conversion period, successful operators are granted organic certification and their goods can be labelled as organic.

6.5.2 Overview of content for EU regulation, US rule and Japanese Agricultural Standard

The EU Council Regulation 2092/91 (from January 2009 substituted by Reg. 834/2007), the USDA NOP and the JAS of Organic Agricultural Products all cover crop production, livestock and processing and handling of organic products. All three regulations include provisions regarding wild harvesting. The EU covers mushrooms and beekeeping, while Japan and the USA do not. The Japanese standard does not cover alcoholic beverages such as wines, while the USA does and the EU is committed to setting specific rules. The USA exempts producers and handlers with less than $5000/year total organic sales from certification requirements, although they must comply with the regulation. The EU and Japan do not allow such an exemption. The EU regulates not only the term 'organic' (or equivalent in other EU languages) but also any other terms that suggest that the product has been produced organically. The US and Japan regulate only the term 'organic' or Japanese equivalents. The format of the EU and Japanese regulations are somewhat similar, resembling the Codex guidelines. This is partly a result of the Japanese basing their regulation on Codex and Codex being heavily influenced by the EU Regulation. The USA regulation follows a different format. All three regulations contain provisions for approval of private certification bodies in implementing the law and provisions for enabling imports from other countries.

6.6 Codex Alimentarius organic norm

The Codex Alimentarius Commission at its 23rd Session in 1999 adopted the Guidelines for the Production, Processing, Labelling and Marketing of Organically Produced Foods, with the exception of the provisions for livestock and livestock products (Codex Alimentarius, 1999). The Codex Alimentarius Commission at its 24th Session in 2001 adopted the sections concerning livestock and livestock products and beekeeping and bee products for inclusion in the Guidelines (Codex Alimentarius Commission, 2001). The main sections of the Guidelines establish the framework within which the more detailed standards in the annexes apply. These sections include, inter alia, the specific labelling requirements; the general rules of production and preparation; requirements for inclusion of input materials in the annexes; and criteria for the development of lists of inputs by countries. Several annexes set down the detailed requirements for production, processing and handling of organic products. These include the rules for the management systems for organic crop production, livestock husbandry and processing (Annex 1) and the permitted agricultural and processing inputs (Annex 2). In addition to the standards for production and processing, the Guidelines contain some provisions regarding inspection and certification systems and import control. Codex standards, codes and related texts have received wider Recognition following the conclusion of the WTO Agreement on the Application of Sanitary and Phytosanitary Measures (SPS) and Technical Barriers to Trade (TBT), as Codex was specifically mentioned under SPS and the reference to international standards in the framework of TBT applies to Codex (see Chapter 2). However, the foreword to the guidelines places certain limitations on its role within the arena of international trade:

> These guidelines are at this stage a first step into official international harmonization of the requirements for organic products in terms of production and marketing standards, inspection arrangements and labelling requirements. In this area the experience with the development of such requirements and their implementation is still very limited. Moreover, consumer perception on the organic production method may, in certain detailed but important provisions differ from region to region in

> the world. Therefore, the following is recognized at this stage... the guidelines do not prejudice the implementation of more restrictive arrangements and more detailed rules by member countries in order to maintain consumer credibility and prevent fraudulent practices, and to apply such rules to products from other countries on the basis of equivalency to such more restrictive provisions. (Codex Alimentarius Commission, 1999)

Codex revision procedures are set down in section 8 of the document. A review of the guidelines is expected to be conducted once every four years. The lists of permitted inputs for production and for processing contained in Annex 2 are subject to review every two years. Both governments and recognized international organizations are invited to make proposals on an ongoing basis. The four years revision has not been undertaken as such since 1999, as only the list of permitted inputs and criteria for substances have benefited from further work.

6.7 International Federation of Organic Agriculture Movements (IFOAM) organic norms

The IFOAM Basic Standards for Organic Production and Processing (IBS) were first published in 1980. Since then they have been subject to biennial review and republication. Democratically and internationally adopted, they became a keystone of the organic movement. The most recent edition of the IFOAM Basic Standards was published together with the IFOAM Criteria for Certification Bodies in the 'IFOAM Norms for Organic Production and Processing' (IFOAM, 2005a). These documents are registered with the International Organization for Standardization (ISO) as international standards in the field of organic agriculture. The introduction to the IFOAM Basic Standards states that these standards 'provide a framework for certification bodies and standard setting organizations worldwide to develop their own certification standards and cannot be used for certification on their own. Certification standards should take into account specific local conditions and provide more specific requirements than the IFOAM Basic Standards' (IFOAM, 2005). The IFOAM norms should therefore be considered as standards for standards in the field of organic agriculture and processing. The introduction also makes it clear that the standards are a reflection of the current state of organic production and processing methods. As such, they should be viewed as a work in progress rather than a final statement. The standards in the IBS are derived from the 'Principal Aims of Organic Production and Processing', which are laid out at the beginning of the document. These principles not only form the basis of the IBS but have also been the guiding principles for national regulations and for international norms such as the Codex Alimentarius Guidelines for organically produced foods. The main sections of the IBS deal with standards for crop production, animal husbandry and processing and handling of organic products. The livestock section establishes generic standards for all livestock. The

Box 6.1 Extract from Codex Alimentarius Guidelines for the Production, Processing, Labelling and Marketing of Organically Produced Foods

SECTION 3: LABELLING AND CLAIMS GENERAL PROVISIONS

3.1 Organic products should be labelled in accordance with the Codex General Standard for the Labelling of Prepackaged Foods.

3.2 The labelling and claims of a product specified in Section 1.1(a) may refer to organic production methods only where:

a) such indications show clearly that they relate to a method of agricultural production;
b) the product was produced in accordance with the requirements of Section 4 or imported under the requirements laid down in Section 7;
c) the product was produced or imported by an operator who is subject to the inspection measures laid down in Section 6, and
d) the labelling refers to the name and/or code number of the officially recognized inspection or certification body to which the operator who has carried out the production or the most recent processing operation is subject.

3.3 The labelling and claims of a product specified in paragraph 1.1(b) may refer to organic production methods only where:

a) such indication show clearly that they relate to a method of agricultural production and are linked with the name of the agricultural product in question, unless such indication is clearly given in the list of ingredients;
b) all the ingredients of agricultural origin of the product are, or are derived from, products obtained in accordance with the requirements of Section 4, or imported under the arrangements laid down in Section 7;
c) the product should not contain any ingredient of non-agricultural origin not listed in Annex 2, Table 3;
d) the same ingredients shall not be derived from an organic and nonorganic origin;
e) the product or its ingredients have not been subjected during preparation to treatments involving the use of ionizing radiation or substances not listed in Annex 2, Table 4;
f) the product was prepared or imported by an operator subject to the regular inspection system as set out in Section 6 of these guidelines; and
g) the labelling refers to the name and/or the code number of the official or officially recognized certification body or authority to which the operator who has carried out the most recent preparation operation is subject.

3.4 By way of derogation from paragraph 3.3(b),

- certain ingredients of agricultural origin not satisfying the requirement in that paragraph may be used, within the limit of maximum level of 5% m/m of the total ingredients excluding salt and water in the final product, in the preparation of products as referred to in paragraph 1.1(b);
- where such ingredients of agricultural origin are not available, or in sufficient quantity, in accordance with the requirements of Section 4 of these guidelines;

Box 6.1 (continued)

3.5 Pending further review of the guidelines in accordance with Section 8, Member Countries can consider the following with regard to products referred to in paragraph 1.1(b) marketed in their territory:

- the development of specific labelling provisions for products containing less than 95% ingredients of agricultural ingredients;
- the calculation of the percentages in 3.4 (5%) and in 3.5 (95%) on the basis of the ingredients of agricultural origin (instead of all ingredients excluding only salt and water);
- the marketing of product with in transition/conversion labelling containing more than one ingredient of agricultural origin.

3.6 In developing labelling provisions from products containing less than 95% of organic ingredients in accordance with the paragraph above, member countries may consider the following elements in particular for products containing 95% and 70% of organic ingredients:

a) the product satisfies the requirements of paragraphs 3.3(c), (d) (e), (f) and (g);
b) the indications referring to organic production methods should only appear on the front panel as a reference to the approximate percentage of the total ingredients including additives but excluding salt and water;
c) the ingredients, appear in descending order (mass/mass) in the list of ingredients;
d) indications in the list of ingredients appear in the same colour and with an identical style and size of lettering as other indications in the list of ingredient.

LABELLING OF PRODUCTS IN TRANSITION/CONVERSION TO ORGANIC

3.7 Products of farms in transition to organic production methods may only be labelled as 'transition to organic' after 12 months of production using organic methods providing that:

a) the requirements referred to in paragraphs 3.2 and 3.3 are fully satisfied;
b) the indications referring to transition/conversion do not mislead the purchaser of the product regarding its difference from products obtained from farms and/or farm units which have fully completed the conversion period;
c) such indication take the form of words, such as 'product under conversion to organic farming', or similar words or phrase accepted by the competent authority of the country where the product is marketed, and must appear in a colour, size and style of lettering which is not more prominent than the sales description of the product;
d) foods composed of a single ingredient may be labelled as 'transition to organic' on the principal display panel;
e) the labelling refers to the name and/or the code number of the official or officially approved certification body or authority to which the operator who has carried out the most recent preparation is subject.

Box 6.2 Extract from IFOAM Basic Standard

7 Labelling
7.1 General
General Principle
Organic products are clearly and accurately labelled as organic.

Recommendations
When the full standards requirements have been fulfilled, products should be labelled as 'produce of organic agriculture' or a similar description.
The name and address of the person or company legally responsible for the production or processing of the product should be on the label.
Product labels should identify all ingredients, processing methods, and all additives and processing aids.
Labels should contain advice on how to obtain all additional product information.
All components of additives and processing aids should be declared.
Wild ingredients or products should be declared as such, as well as organic.

Standards shall require that:
7.1.1 The person or company legally responsible for the production or processing of the product and the certification body shall be identifiable.
7.1.2 To be labelled as 'produce of organic agriculture' or equivalent protected terms, a product shall comply with at least these standards.
7.1.3 Mixed products where not all ingredients, including additives, are of organic origin and products that are entirely in compliance with these standards, shall be labelled in the following way (percentages in this section refer to raw material weight):

a. where a minimum of 95% of the ingredients are of certified organic origin, products may be labelled 'certified organic' or equivalent and should carry the certification mark of the certification body;
b. where less than 95% but not less than 70% of the ingredients are of certified organic origin, products may not be called 'organic'. The word 'organic' may be used on the principal display in statements like 'made with organic ingredients' provided there is a clear statement of the proportion of the organic ingredients. An indication that the product is covered by the certification body may be used, close to the indication of proportion of organic ingredients;
c. where less than 70% of the ingredients are of certified organic origin, the indication that an ingredient is organic may appear in the ingredient list. Such product may not be called 'organic.'

7.1.4 All ingredients of a multi-ingredient product shall be listed on the product label in order of their weight percentage. It shall be apparent which ingredients are of organic certified origin and which are not. All additives shall be listed with their full name.
If herbs and/or spices constitute less than 2% of the total weight of the product, they may be listed as 'spices' or 'herbs' without stating the percentage.

Box 6.2 (continued)

7.1.5 Added water and salt shall not be included in the percentage calculations of organic ingredients.
7.1.6 The label for conversion products shall be clearly distinguishable from the label for organic products.
7.1.7 Organic products shall not be labelled as GMO-free in the context of these standards. Any reference to genetic engineering on product labels shall be limited to the production and processing methods themselves having not used GMOs.

exception is beekeeping which is dealt with in a separate section. Additional sections of the standards set out the requirements for ecosystems, labelling and social justice. Each section of the IBS is presented as General Principles, Recommendations and Standards. The General Principles are the goals that organic production and processing work towards. The Recommendations provide standards that IFOAM promotes but does not require. The Standards are the minimum requirements that must be fully incorporated into certification standards.

6.8 Private standards

The Soil Association in the UK published the first private organic standards in 1967 followed by Nature et Progrès, France 1972. These were more a set of guiding principles rather than the detailed production and processing standards prevalent today. It is important to realize that this initiative, and other private standards that were developed in Europe, the USA and elsewhere shortly thereafter, was driven by the need of organic farmers in the region to have a common definition of organic. This was both to provide assurance to the growing consumer

Fig. 6.2 Private organic logo. Soil Association, UK. Reproduced with permission of the Soil Association Organisation.

Fig. 6.3 Private organic logo. KRAV private logo Sweden. Reproduced with the permission of KRAV Ekonomisk förening (Sweden).

sector and to prevent fraudulent claims and unfair competition. Farmers' associations published all of the earliest organic standards. Along with publishing standards, the association then set about verifying compliance with those standards. The result was that certification bodies that were established during the 1970s and 1980s also published their own standards. These standards provided an identity to the farmers' association and helped to ensure the loyalty of the farmer. The result of this heritage is that there are a great many private organic standards for production and certification around the globe. This plethora of standards has created some difficulties with respect to mutual recognition and trade; there have also been some advantages. As the standards are being set in the specific region in which the certification body operates, they tend to be more appropriate to the local ecosystems and local culture than standards set distantly. It is noteworthy that private organic standards have been developed for activities generally not covered in regulations. These included animal husbandry before the regulations were adopted, textile processing, aquaculture, forestry and others. The private standards determined the content of the IFOAM Basic Standards which, in turn, have had a major influence on the EU Regulation 2092/91, which itself has influenced the content of most other organic regulations and the Codex Alimentarius Guidelines.

6.8.1 Labelling and certification as a marketing tool

From the early stages in the development of organic certification the private certification bodies have marketed their certification symbols to the consumer as a guarantee of quality. The degree to which they have been successful differs from country to country. In some countries such as Sweden, Switzerland and the UK there is strong consumer identification with the certification body's symbol, whereas in other countries such as the USA there is little consumer recognition of the symbols. The certification bodies' symbols are generally officially registered as trademarks. IFOAM, the EU, the USA and Japan all allow use of their approvals on packaging.

6.9 European Union organic regulation

In 1991, the European Council of Agricultural Ministers adopted Regulation (EEC) No. 2092/91 on organic farming and the corresponding labelling of agricultural products and foods. The introduction of this Regulation was part of the reform of the EU Common Agricultural Policy and represented the conclusion of a process through which organic agriculture received the official recognition of the 15 states which were EU members at the time. At the beginning, the organic Regulation only regulated plant products. In 1999 additional provisions for animal products were introduced (EEC, 1999). These rules included animal feed, prevention of illness, veterinary treatment, animal protection, livestock breeding in general and the use of livestock manure. In the same regulation (EC 1804/99) the use of genetically modified organisms and products produced from them was expressly excluded from organic production. At the same time, the import of organic products was approved from third countries whose production criteria and systems of control could be recognized as equivalent to those of the EU. As a result of this ongoing process of supplementation and amendment, the provisions contained in Regulation (EEC) No. 2092/91 became very complex and extensive. The level of importance that the original EU organic Regulation enjoyed lay in the fact that it created common minimum standards for the entire EU. In this process, the confidence of consumers, who could purchase organic products from other member states with the certainty that these products fulfilled the same minimum requirements, was strengthened. It was left up to the member states and private organizations to enact their own additional stricter standards.

6.9.1 The new EU Regulation

In June 2007, the European Council of Agricultural Ministers agreed to a new Council Regulation on organic production and labelling of organic products (Reg. 834/07). This new Council Regulation contains clearly defined goals, principles and general rules for organic production. On 1 January 2009, new EU regulations went into effect for the production, control and labelling of organic products. However, some of the new provisions on labelling do not take effect until 1 July 2010. Since the EU now includes 27 member states, and extends from the far north to southern and eastern Europe, local climatic, cultural or structural differences can be compensated through foreseen flexibility rules. Foods may only be marked as 'organic' if at least 95% of their agricultural ingredients are organic. The '70–95%' organic ingredients category of the previous Regulation disappeared. Organic ingredients in non-organic food may be considered as organic in the list of ingredients, as long as this food has been produced in accordance with the organic legislation. In order to ensure better transparency, the code number of the control body must be indicated. The use of genetically modified organisms (GMO) and of products manufactured from GMOs is still prohibited in organic production. Products containing GMOs may not be labelled as organic unless the ingredients containing GMOs entered the products unintentionally and the GMO proportion in the ingredient is less than 0.9% (see Chapter 10).

The community logo
The EU organic logo and those of EU Member States are used to supplement the labelling and increase the visibility of organic food and beverages for consumers.

Products bearing the EU logo have to fulfil the following conditions:

- at least 95% of the product's ingredients of agricultural origin have been organically produced;
- the product complies with the rules of the official inspection scheme;
- the product has come directly from the producer or preparer in a sealed package;
- the product bears the name of the producer, the preparer or vendor and the name or code of the inspection body.

The placement of the EU logo is currently voluntary, but will become mandatory as of 1 July 2010 for pre-packaged food. It will continue to be voluntary for imported products after this date. From 1 July 2010, where the Community logo is used, an indication of the place where the agricultural raw materials were farmed should accompany it. It should be indicated that the raw materials originate from 'EU Agriculture', 'non-EU Agriculture' or 'EU/non-EU Agriculture'. If all raw materials have been farmed in one country, the name of this specific country, inside or outside the EU, can be indicated instead. If operators wish to sell their products in an EU Member State other than their own, they may place an additional national or private logo that will be recognized by the consumers of this particular country (EC, 2008).

Organic import
The distribution of organic products from third countries is only permitted in the common market, when they are produced and controlled under the same or equivalent conditions. The import regime has been expanded with the new legislation. The procedure for import licences will in the future be replaced by a new import regime. Control bodies working in third countries will then be directly authorized and monitored by the European Commission and the Member States. This new procedure allows the EU Commission to supervise and monitor the import of organic products and the control of the organic guarantees.

New fields of application
A basis for the acceptance of EU rules on organic aquaculture and seaweeds was established in the new legislation.

Indications
The Regulation contains clear and strict rules about labelling and logo use, to minimize any confusion among consumers, or potential misuse: 'Any terms such as organic, bio, eco, etc., including terms used in trademarks, or practices used in labelling or advertising liable to mislead the consumer or user by suggesting that a product or its ingredients satisfy the requirements set out under this Regulation shall not be used for non-organic products' (EC, 2007). In addition, the organic label cannot be used for a product that contains GMOs. To provide further confidence, by law all products labelled as organic must bear the name of the last

operator who has handled the product, e.g. the producer, the processor or the distributor and the name or code number of their inspection body.

Specific labelling requirements
The Commission is committed to establish specific labelling and composition requirements applicable to:

- organic feed;
- in conversion products of plant origin;
- vegetative propagating material and seeds for cultivation.

6.10 United States organic rules (National Organic Program)

The Organic Foods Production Act (OFPA) and NOP as implemented in 2002 offer assurance that all food products labelled as organic in the USA are governed by consistent standards. The labelling requirements of the NOP apply to raw, fresh products and processed products that contain organic agricultural ingredients. Agricultural products that are sold, labelled or represented as organic must be produced and processed in accordance with the NOP standards. Except for operations whose gross income from organic sales totals $5000 or less, farm and processing operations that grow and process organic agricultural products must be certified by USDA accredited certifying agents.

6.10.1 NOP labelling

Labelling requirements are based on the percentage of organic ingredients in a product, allowing four different labelling options based on the percentage of organic ingredients in a product. These include three distinct categories, and a fourth option for products that contain organic ingredients but not at a high enough level to meet one of the three labelling categories:

- *100 percent organic*: Only products that have been exclusively produced using organic methods and that contain only organic ingredients (excluding water and salt) are allowed to carry a label declaring '100 percent organic.'
- *Organic:* This signifies that at least 95 percent of the ingredients (by weight, excluding water and salt) in a processed product have been organically produced. The remaining contents can only be natural or synthetic ingredients not available in an organic form that are recommended by the National Organic Standards Board and allowed on the National List. The product cannot use both organic and non-organic versions of any ingredient that is listed as organic. For instance, if a loaf of bread is made with organic wheat, all of the wheat in the bread must be organic.
- *Made with organic*: Products with 70–95 percent organic ingredients may display 'Made with organic [with up to three specific organic ingredients or food groups listed]' on the front panel.

All three categories prohibit the inclusion of any ingredients produced using

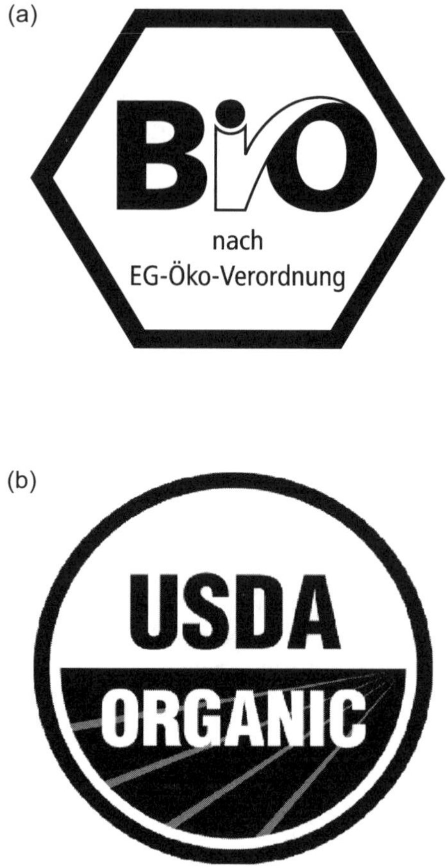

Fig. 6.4 Public organic logos. (a) BIO label from Germany. Reproduced with the permission of Bio-Siegel, Federal Ministry of Food, Agriculture and Consumer Protection, Germany. (b) US organic seal. Reproduced by permission of the National Organic Program, United States Department of Agriculture.

genetic engineering, irradiation, or sewage sludge. Products with less than 70 percent organic ingredients can list the organic items only in the ingredient panel. There can be no mention of organic on the main panel.

US Department of Agriculture organic logo

To assist consumers, USDA has designed a seal that may be used only on products labelled as '100 percent organic' or 'organic.' Use of the seal is voluntary, but is seen as a useful tool. Grocery stores are increasingly using the 'USDA Organic' seal on shelf talkers and other point of purchase materials to help identify organic sections in the store. Non-food products that meet the requirements for using the 'USDA Organic' seal can also use the seal.

6.11 Japanese Agricultural Standard (JAS) of Organic Agricultural Products

6.11.1 JAS standards

'JAS' is the abbreviation for 'Japanese Agricultural Standard,' and is currently used as a term that represents the overall system. The Japanese Agricultural Standards applied to individual commodities are referred to as 'JAS Standards.' The JAS System was introduced in 1950 as the Agricultural and Forestry Standard Law, and assumed its current status in 1970 with the addition of a quality labelling standard system. At present, the JAS System consists of the combination of the 'JAS Standards System' and the 'Quality Labelling Standards System.' JAS Standards are established for types of agricultural and forestry products designated by the Minister of Agriculture, Forestry and Fisheries. Establishment and other procedures must follow resolutions by the 'Council for Agricultural and Forestry Standard' (JAS Council), a body comprising consumers, producers, commercial users, academic experts and others.

6.11.2 JAS organic standards

The JAS Standards for organic plants and organic processed foods of plant origin were established in 2000 on the basis of the Guidelines for the Production, Processing, Labelling and Marketing of Organically Produced Foods which were adopted by the Codex Alimentarius Commission. The organic JAS system has been further developed with the addition of the JAS Standards for organic livestock products, organic processed foods of animal origin and organic feeds which took effect in November 2005. Operators certified by registered Japanese or overseas certifying bodies are able to attach the organic JAS logo to products that were produced or manufactured in accordance with relevant organic JAS Standards.

6.12 International Task Force for organic regulations harmonization and equivalence

To face the growing complexities and challenges derived from a non-harmonized regulatory system on organic agriculture, a number of individuals working in government agencies, inter-governmental agencies and civil society and other private sector organizations involved in organic agriculture regulation, standardization, accreditation, certification and trade joined together from 2003 to 2008 in a platform for dialogue among public and private stakeholders: International Task Force for organic regulations harmonization and equivalence (ITF). The goal of the ITF was to address and seek solutions to trade barriers arising from the many different standards, technical regulations and certification requirements that function in the organic sector, and enable developing countries to have more access to organic trade. Jointly led by FAO, IFOAM and UNCTAD, the ITF focused on oppor-tunities for harmonization, equivalence, recognition and other forms of cooperation within and between government and private organic guarantee systems. Its formal results include technical studies and briefing papers, recommendations and tools for solutions. ITF produced two practical tools for harmonization and equivalence:

- The International Requirements for Organic Certification Bodies (IROCB), a reference norm that can be used by governments and private accreditation and certification bodies as a means of accepting certification of organic products outside of their own system (ITF, 2008a).
- The Guide for Assessing Equivalence of Organic Standards and Technical Regulations (EquiTool), a set of procedures and criteria for deciding when a standard applicable in one region of the world is equivalent to a standard applicable in another region (ITF, 2008b).

The ITF agreed to support the two international standards for organic production and processing (IFOAM standards in the private sector and Codex Alimentarius Commission standards in the government sector), and encourage harmonization and equivalence based on these standards.

6.13 References

CODEX ALIMENTARIUS COMMISSION (1999) Guidelines for the Production, Processing, Labelling and Marketing of Organically Produced Foods (Except sections for livestock production) ALINORM 99/22, Appendix II and ALINORM 99/22A, para. 33 and Appendix II Adopted by the Codex Alimentarius Commission Twenty-Third Session, FAO Headquarters, Rome, 28 June–3 July 1999. Rome: FAO/WHO.

CODEX ALIMENTARIUS COMMISSION (2001) Codex Guidelines for the Production, Processing, Labelling and Marketing of Organically Produced Foods. Provisions concerning Livestock Production ALINORM 01/22, Appendix II. Adopted by Codex Alimentarius Commission Twenty-fourth Session, Geneva, Switzerland, 2–7 July 2001. Rome: FAO/WHO.

EC (1992) European Community Council Regulation (EEC) No. 2078/92 on agricultural production methods compatible with the requirements of the protection of the environment and the maintenance of the countryside of 30 June 1992. CELEX-EUR. *Official Journal of the European Union*, L 215, 30 July, pp. 85–90.

EC (2007) Council Regulation (EC) No 834/2007 of 28 June 2007 on organic production and labelling of organic products and repealing Regulation (EEC) No 2092/91. *Official Journal of the European Union*, L189, 20 July, pp. 1–23.

EC (2008) Commission Regulation (EC) No 889/2008 of 5th September 2008 laying down detailed rules for the implementation of Council Regulation (EC) No 834/2007 on organic production and labelling of organic products with regard to organic production, labelling and control. *Official Journal of the European Union*, L250, 18th September, pp. 1–84.

EEC (1991) EEC Council Regulation 2092/91 of 24 June 1991 on organic production of agricultural products and indications referring thereto on agricultural products and foodstuffs. *Official Journal of the European Union*, L198, 22 July, pp. 1–15.

EEC (1999) Council Regulation (EC) No 1804/1999 of 19 July 1999 supplementing Regulation (EEC) No 2092/91 on organic production of agricultural products and indications referring thereto on agricultural products and foodstuffs to include livestock production. *Official Journal of the European Union*, L222, 24 August, pp.1–28.

IFOAM (2005a) *The IFOAM Norms Version 2005*. Bonn: International Federation of Organic Agriculture Movements.

IFOAM (2005b) *Organic Agriculture Principles*. Bonn: International Federation of Organic Agriculture Movements. Available at: http://www.ifoam.org/about_ifoam/principles/index.html [accessed July 2009].

IFOAM (2008) *Definition of Organic Agriculture*. Bonn: International Federation of Organic Agriculture Movements. Available at: http://www.ifoam.org/growing_organic/definitions/doa/index.html [accessed July 2009].

ITF (INTERNATIONAL TASK FORCE FOR ORGANIC REGULATIONS HARMONIZATION AND EQUIVALENCE) (2008a) *International Requirements for Organic Certification Bodies*. Geneva: UNCTAD. Available at: http://r0.unctad.org/trade_env/itf-organic/meetings/itf8/IROCB_0809%20.pdf [accessed July 2009].

ITF (INTERNATIONAL TASK FORCE FOR ORGANIC REGULATIONS HARMONIZATION AND EQUIVALENCE) (2008b) *EquiTool*. Geneva: UNCTAD. Available at: http://r0.unctad.org/trade_env/itf-organic/meetings/itf8/ITF_EquiTool_finaldraft_080915db2.pdf [accessed July 2009].

JAPAN MINISTRY OF AGRICULTURE, FORESTRY AND FISHERIES (2000) Japanese Agricultural Standard JAS of Organic Agricultural Products Notification No.59 of the Ministry of Agriculture, Forestry and Fisheries of January 20, 2000. Tokyo.

SPC (2008) *Pacific Organic Standard*. Nabua, Fiji: Secretariat of the Pacific Communities.

UNITED STATES CONGRESS (1990) United States Public Law 1990. The Organic Foods Production Act. P.L. 101–624, Title 21. Codified at 7 United States Code 6501 *et seq*. Washington, DC.

UNITED STATES DEPARTMENT OF AGRICULTURE (2000) National Organic Program; Final Rule. 7 CFR Part 205. National Archives and Records Administration Office of the Federal Register 65 80548. Washington, DC.

WILLER, H. AND KILCHER, L. (EDS) (2009) *The World of Organic Agriculture – Statistics and Emerging Trends 2009*. IFOAM, Bonn; FiBL, Frick; ITC, Genf (www.organic-world.net) [accessed 27th March 2009].

7

FAO's ecolabelling guidelines for marine capture fisheries: an international standard

Rolf Willmann, Kevern Cochrane and William Emerson, Food and Agriculture Organization of the United Nations (FAO), Italy

Abstract: Many of the world's marine fisher resources are either overfished or fully exploited and global production from wild stocks is close to its long run biological maximum. Consumer awareness about the serious condition of many marine fishery resources has grown, especially in OECD countries. The objective of eco-labelling of fish and fishery products is to achieve the goal of sustainable fisheries, in line with the FAO Code of Conduct for Responsible Fisheries and other related international instruments.

Key words: fisheries ecolabelling, FAO guidelines, equivalence, barriers to trade.

7.1 Introduction

Many of the world's marine fisher resources are either overfished or fully exploited. In 2008, the Food and Agriculture Organization's *State of World Fisheries and Aquaculture* (SOFIA) reported that more than one quarter of the monitored fish stocks were overfished, depleted or recovering while another more than one half were fully exploited, which means that they were estimated to be producing catches at the maximum that could be sustained over time. In a poorly managed fishery, full exploitation may just be an intermediate state of a stock on its way to being over-exploited. Only one-fifth of the stocks – down from two-fifths in the 1970s – remain under-exploited or moderately exploited. Global production of seafood from wild stocks is therefore close to its long run biological maximum (FAO 2008a).

The world's marine fisheries are also performing badly in economic terms. A recent World Bank–FAO study notes that the contribution of the harvest sector of the world's marine fisheries to the global economy is substantially smaller than it could be. It estimates the annually lost economic benefits in the order of $50 billion. Over the last three decades, the cumulative global loss of potential economic benefits is estimated in the order of $2 trillion. The losses represent the difference between the potential and actual net economic benefits from global marine fisheries (World Bank and FAO 2009).

Those who carry the heaviest burden of over-exploited fishery resources in biological and economic terms are the millions of often poor and vulnerable fishery-dependent communities of developing countries and low income consumers who rely on fisheries and fish for their livelihoods and food security.

7.2 Why ecolabelling

The objective of ecolabelling of fish and fishery products is to achieve the goal of sustainable fisheries. This is in line with the objectives pursued through the FAO Code of Conduct for Responsible Fisheries (CCRF) and other related international instruments, in particular the 1982 United Nations Convention on the Law of the Sea and the 1995 Agreement for the Implementation of the Provisions of the United Nations Convention on the Law of the Sea of 10 December 1982 relating to the Conservation and Management of Straddling Fish Stocks and Highly Migratory Fish Stocks.[1]

Ecolabelling is a market-based instrument which usually relies on and reinforces management measures taken by government fisheries management agencies.[2] The Nordic Technical Working Group on Eco-labelling Criteria (2000) identified the following positive incentives that are created by ecolabels for products from capture fisheries:

- The fishing community is provided with a market incentive to request that authorities manage fish stocks in a responsible precautionary way.
- Governments are given an incentive to upgrade their fisheries management practices to improve the market situation for national fisheries products.
- Authorities are given an incentive to improve research and the monitoring of their fish stocks and fisheries.

Consumers' product choices and their willingness to pay a higher price for an ecolabelled product will depend on their general responsiveness and capacity to

[1]The text of the 1982 Convention and 1995 Agreement can be found on this Internet address: http://www.un.org/Depts/los/convention_agreements/texts/unclos/closindx.htm

[2]A comprehensive review of the principles and practice of seafood ecolabelling has been edited by Ward and Phillips (2008).

address environmental concerns through their purchasing behaviour and their awareness and understanding of the specific objectives pursued through the labelling scheme. Consumer awareness about the serious condition of many of the world's marine fishery resources has grown globally but especially in OECD countries which include many of the major importers of seafood products. Fishery products are among the most traded and valued food products. In 2006, nearly 40 percent of global fish production was internationally traded at an aggregate value of $85.9 billion of which nearly one half by developing countries (FAO, 2008a).

There are growing numbers of consumers in Japan, Germany, Switzerland, the UK, the USA, and other countries, including urban consumers of developing countries, who take into account the environmental impact of their purchases including fishery resources. For some years, there have been ecolabelled seafood products from the Marine Stewardship Council (MSC) or following the numerous and varied recommended buyers' lists from environmental organisations of 'sustainable' or 'non-sustainable' fish purchases, although some are not fully reliable. Consumers are increasingly lured by many of the major retail chains which have guidelines for their suppliers regarding various criteria including environmental friendliness as well as employee working conditions.

7.3 History of the FAO ecolabelling guidelines for fish and fishery products

The impetus for addressing the issue of ecolabelling of fisheries products in FAO arose from the launch of the MSC initiative by Unilever PLC/NV and WWF, a leading environmental organisation, in early 1996. In their joint Statement of Intent, WWF noted its wish for 'a new approach to ensure more effective management of marine life', while Unilever PLC/NV, a major buyer of frozen fish and manufacturer of many of the world's best-known frozen-fish products, expressed 'its commitment to long-term fish stock sustainability to ensure a future for its successful fish business'.[3]

The reactions to the initiative of WWF and Unilever were mixed. While it was applauded by some industry groups, conservation organisations and governments, many fisheries stakeholders and governments were initially sceptical about the intentions of this unlikely partnership between a big corporate player in the fish processing and retailing business and an environmental non-governmental organisation (NGO) which until that time was perceived as having greater interests in marine conservation than supporting the fishing industry.

At the intergovernmental level, the matter was discussed controversially in several sessions of the FAO Committee on Fisheries (COFI) and sessions of its Sub-Committee on Fish Trade from 1997 onwards. It was also considered in a technical consultation of government-nominated experts in 1998 which investigated the feasibility and practicability of developing non-discriminatory, globally

[3]Cited in ICSF (1998; p. 6).

applicable, technical guidelines for the ecolabelling of fish and fishery products from marine fisheries. Unanimity among FAO members on the need for an international normative instrument on fisheries ecolabelling was, however, only reached at COFI, 2003.

Several factors are likely to have influenced a change in COFI, 2003 by relatively few countries which had opposed international ecolabelling guidelines. The issue of labelling requirements for environmental purposes had become, since the 4th World Trade Organization (WTO) Ministerial Conference in Doha, November 2001, an issue of special focus in the work of the WTO Committee on Trade and Environment (CTE). At Doha, WTO members instructed the CTE to undertake further work on labelling requirements for environmental purposes and in particular to:

- look at the impact of ecolabelling on trade,
- examine whether existing WTO rules stood in the way of ecolabelling policies, and
- identify any WTO rule that would need to be clarified.

In its report to the 5th Session of the WTO Ministerial Conference in Cancún, most CTE Members agreed that voluntary, participatory, market-based and transparent environmental labelling schemes were potentially efficient economic instruments that informed consumers about environmentally friendly products. Importantly, the report noted that ecolabelling tended, generally, to be less trade restrictive than other instruments. However, it also noted that environmental labelling schemes could be misused for the protection of domestic markets. Hence, these schemes needed to be non-discriminatory and not result in unnecessary barriers or disguised restrictions on international trade (WTO, 2003).

Another important factor that might have tipped the balance in favour of the development of FAO ecolabelling guidelines was the fact that the MSC programme was moving successfully ahead and encompassing an increasing number of fisheries and certified product lines. There was also an increasing number of large wholesale and retail chains which announced green procurement guidelines for their fishery products and commitments in the medium term to only procure fish from sustainable sources, including MSC certified fisheries. Thus, it became clear that important segments of market demand in the large fish importing countries were moving towards certified products. A 'green image' became an important strategy to maintain and expand market shares in the food products industry.

Thus a consensus emerged among FAO members on the need for international harmonisation of criteria and procedures and related issues such as equivalence and mutual recognition. This would avoid ecolabelling programmes in fisheries discriminating against certain producers, kinds of fisheries or countries. It would also help to avert a situation that may arise where a series of competing ecolabelling schemes were to apply different standards and criteria causing confusion rather than providing for more informed purchasing behaviour by consumers (Cochrane and Willmann, 2000).

7.4 The development of the FAO ecolabelling guidelines

With the blessing of its members provided at COFI, 2003, FAO initiated the process of developing international ecolabelling guidelines for fish and fishery products from marine capture fisheries. FAO first convened a consultation of experts in their individual capacities in October 2003. The Consultation brought together experts from different disciplines, regions and institutional backgrounds (government, industry, conservation organisations, small-scale fishers) of whom several took part in subsequent negotiation stages as members of their government delegations or as non-governmental observers. The report from the Expert Consultation (FAO, 2003) provided a background document for the subsequent Technical Consultation. The Technical Consultation of experts representing their governments and observer organisations initiated the intergovernmental negotiations proper, in October 2004 (FAO, 2005a). As number of issues of concern and controversy could not be resolved during that meeting, another round of consultation was held to try to reach agreement on these just prior to the 26th session of COFI in March 2005. A particular concern, particularly amongst developing countries, was, and still is, that the ecolabels could be used as technical barriers to trade. Negotiations continued alongside COFI in a small group representing the different regions and interest groups. COFI adopted the final text by consensus, but a few country delegations expressed reservations that have been reflected in the COFI Report (FAO, 2005b). Subsequently, the FAO Secretariat was asked to undertake further work on the minimum substantive requirements set out in the guidelines. After consultation with a group of experts, some amendments to and expansions of this section of the 2005 guidelines were proposed (FAO, 2008b). These amendments were adopted by the 2009 Session of COFI in March 2009 and the revised final guidelines will be published this year.[4]

In the following section, selected provisions of the guidelines including the latest revisions are presented to explain the key intent of the guidelines and comment on the evolution of the text through the various drafting and negotiation stages. Their normative basis is indicated in the guidelines themselves and includes, in particular, the 1982 UN Convention on the Law of the Sea, the 1995 UN Fish Stocks Agreement, the 1995 Code of Conduct for Responsible Fisheries, relevant guides of the International Organization for Standardization (ISO) and provisions of the WTO Technical Agreement on Barriers to Trade, especially ANNEX 3 *Code of Good Practice for the Preparation, Adoption and Application of Standards.*

The text of this chapter follows the structure of the guidelines – Scope, Principles, General considerations, Terms and definitions, Minimum substantive requirements and criteria and Procedural and institutional aspects.

---

[4]The text of the 2008 amendments to the 2005 Ecolabelling Guidelines is shown in Appendix E of FAO (2008b): ftp://ftp.fao.org/docrep/fao/010/i0006e/i0006e00.pdf. The text of the 2005 Guidelines is shown here: ftp://ftp.fao.org/docrep/fao/008/a0116t/a0116t00.pdf)

7.4.1 Scope

The initial mandate by COFI in 2003 was to develop guidelines for the ecolabelling of marine capture fisheries only. This was extended to inland capture fisheries, as a separate set of guidelines, by COFI in 2005. Draft ecolabelling guidelines for inland fisheries were developed by an expert consultation in 2006, but further work was requested by COFI, 2007 and 2009. As they are not yet finalised, they are not cited in the following. However, key differences between the two sets of guidelines are explained in the annotations.

The text of Paragraph 1 on the scope reads as follows:

These guidelines are applicable to ecolabelling schemes that are designed to certify and promote labels for products from well-managed marine capture fisheries and focus on issues related to the sustainable use of fisheries resources.

The exclusion of social and economic or health and safety aspects from the scope of the guidelines is notable. This is in keeping with the views expressed by most governments at the 1998 FAO technical consultation. It would have been very unlikely, if at all possible, that international consensus among governments could have been reached on standards relating to social and economic factors.

As a significant portion of production from inland waters is derived from culture-based and enhanced fisheries these fisheries, are included within the scope of inland capture fisheries although the actual boundary between aquaculture and capture-based fisheries for the purposes of ecolabelling guidelines still needs to be clarified (FAO, 2006, 2008b).

7.4.2 Principles

The full set of principles elaborated by the 1998 technical consultation were maintained with some amendments and re-ordering. For the inland fisheries guidelines, references to the Convention on Biodiversity and the Ramsar Convention on Wetlands were added.

The text of the Principles reads:

2. *The following principles should apply to ecolabelling schemes for marine capture fisheries:*

2.1 *Be consistent with the 1982 United Nations Convention on the Law of the Sea and the Agreement for the Implementation of the Provisions of the United Nations Convention on the Law of the Sea of 10 December 1982 relating to the Conservation and Management of Straddling Fish Stocks and Highly Migratory Fish Stocks, the FAO Code of Conduct for Responsible Fisheries and the World Trade Organization (WTO) rules and other relevant international instruments.*

2.2 *Recognize the sovereign rights of States and comply with all relevant laws and regulations.*

2.3 *Be of a voluntary nature and market-driven.*
2.4 *Be transparent, including balanced and fair participation by all interested parties.*
2.5 *Be non-discriminatory, do not create unnecessary obstacles to trade and allow for fair trade and competition.*
2.6 *Provide the opportunity to enter international markets.*
2.7 *Establish clear accountability for the owners of schemes and the certification bodies in conformity with international standards.*
2.8 *Incorporate reliable, independent auditing and verification procedures.*
2.9 *Be considered equivalent if consistent with these guidelines.*
2.10 *Be based on the best scientific evidence available, also taking into account traditional knowledge of the resources provided that its validity can be objectively verified.*
2.11 *Be practical, viable and verifiable.*
2.12 *Ensure that labels communicate truthful information.*
2.13 *Provide for clarity.*
2.14 *Be based, at a minimum, on the minimum substantive requirements, criteria and procedures outlined in these guidelines.*
3. *The principle of transparency should apply to all aspects of an ecolabelling scheme including its organizational structure and financial arrangements.*

Principle 2.1 was included to give assurance that ecolabelling schemes will not contravene widely accepted or ratified international instruments.

Principle 2.14 establishes the FAO guidelines as a minimum standard for any capture fisheries ecolabelling scheme. Whether the guidelines should be considered *the* international standard within the framework of WTO rules and regulations is open to interpretation for reasons discussed in greater detail in Wessells *et al.* (2001)· WTO does not claim to be the appropriate forum for discussions on the general usefulness of ecolabelling schemes or what constitutes appropriate criteria for assessing sustainability. Indeed, WTO explicitly defers such issues to international agreements or bodies with appropriate expertise (Wessells *et al.*, 2001).

The precise formulation of Principle 2.6 was controversial until nearly the end of the negotiations that took place during the process of adoption of the guidelines in COFI, 2005. Some felt that its intent was already reflected in Principles 2.1 and 2.5. Others argued that the idea of gaining better access to international markets through ecolabelling schemes should be a principle by itself. Whereas Principle 2.1 already refers to consistency with WTO rules, a reiteration of this requirement was felt necessary for Principles 2.5 and 2.6. This emphasis needs to be interpreted in the context of the debate in WTO, in particular within its Committee on Trade and the Environment, on the applicability of WTO rules to environmental labelling (see Wessells *et al.*, 2001). In relation to this subject, it needs to be recalled that the tuna–dolphin and shrimp–turtle trade disputes have likely caused sensitivities among several countries and a cautious attitude to the potential implications of international ecolabelling guidelines should a trade dispute arise.[5]

7.4.3 General considerations

The intention of this section is to create, to the extent possible, an equal playing field among countries by, *inter alia*, recognising the special conditions and requirements of developing countries and countries in transition on the one hand, while calling for one unique minimum standard on the other hand, in order to avoid any notion of superior or inferior categories of ecolabelled fish and fishery products.

The section also addresses the view of many governments that they should be fully involved, not just individually but also as members of Regional Fisheries Management Organizations (RFMOs) in ecolabelling schemes.[6] It recognises that governments play, or need to play, a paramount and often indispensable role in fisheries management.

In the strict sense, RFMOs do not exist for inland capture fisheries, but the inland fisheries experts agreed to adopt throughout the text of the guidelines the wider term of regional fishery body (RFB) applicable to both RFMOs as well as to bodies having purely advisory functions.

4. *Ecolabelling schemes should take into account that principles, minimum substantive requirements, criteria and procedures set out in this document will apply equally for developed, transition and developing countries.*
5. *Bearing in mind that ecolabelling schemes relate to fisheries management, and rights and duties of States, it is recognized that the involvement of States in ecolabelling schemes is desirable and should be encouraged. It is also recognized that States and, as appropriate, Regional Fisheries Management Organizations (RFMOs) may develop ecolabelling schemes in a manner consistent with these guidelines. Ecolabelling schemes should give full consideration to the recommendations and advice by States, and, as appropriate, RFMOs.*
6. *In accordance with Article 5 of the Code of Conduct for Responsible Fisheries, and recognizing that all countries should have the same opportunities, and in view of the special*
7. *conditions applying to developing countries and countries in transition and their important contribution to international fish*

---

[5]Information on these trade disputes is available on the WTO Internet site at: http://www.wto.org/english/tratop_e/envir_e/edis04_e.htm and http://www.wto.org/english/tratop_e/envir_e/edis08_e.htm

[6]RFMOs are intergovernmental fisheries organisations or arrangements which have the competence to establish fisheries conservation and management measures that are binding on their members. They are the principal mechanism for cooperation between and among coastal states and fishing nations for the management of international fisheries.

trade, it is acknowledged that in order to benefit from applying ecolabelling schemes, states, relevant intergovernmental and non-governmental organizations and financial institutions should provide developing countries and countries in transition with financial and technical assistance to develop and maintain appropriate management arrangements that will allow them to participate in such schemes. Such assistance should also consider direct support towards the often high costs of accreditation and certification. Development agencies and donor institutions are encouraged to support FAO in facilitating financial and technical assistance to developing countries and countries in transition.

7.4.4 Terms and definitions

The section draws heavily on terminology, definitions and standards agreed within the framework of the International Organization for Standardization (ISO) dealing with general requirements on accreditation and certification. It also contains a series of definitions that were specifically developed by the expert and technical consultations for the marine and inland capture fisheries ecolabelling guidelines.

The concept of the unit of certification (paragraph 25) is of special interest as it provides for the possibility of a fishery becoming certified which harvests only a component of a stock. As will become evident in the next section, for purposes of gauging the health of the stock, however, the impact of all fisheries on this stock would have to be taken into account.

The inland fisheries experts concluded that geographic boundaries of inland fisheries did not need to be defined. 'As fish stocks contributing to river, lake and reservoir fisheries may also, in some cases, be caught in estuarine and marine areas, the consideration of impacts of all fisheries utilizing a stock or stocks across their entire area of distribution, including all life stages, is an important element of assessing the state of the "stock under consideration" ' (FAO, 2006).

The expert consultation on inland fisheries added terms for culture-based fisheries, enhanced fisheries and introduced species. The experts drew a line between capture fisheries and aquaculture that permits artificial stocking but not artificial feeding. Whereas the 2006 expert group meeting concluded that the enhancement features of many inland fisheries are the critical distinction from marine capture fisheries, a more recent expert consultation convened by FAO in March 2008 concluded that enhancements are increasingly used too in marine fisheries. It noted that there is no agreed boundary to determine when a fishery applying enhancement measures should cease to be considered a capture fishery. Thus the 2008 group of experts was not in full agreement on the validity of the definitions provided by the 2006 consultation and recommended that additional work be undertaken on these definitions (FAO, 2008b).

7. *For the purpose of these International Guidelines, the following terms and definitions apply.*

<u>*Accreditation*</u>

8. Procedure by which a competent authority gives formal recognition that a qualified body or person is competent to carry out specific tasks.
(Based on ISO/IEC Guide 2:1996, 12.11)

<u>*Accreditation body*</u>

9. Body that conducts and administers an accreditation system and grants accreditation.
(Based on ISO Guide 2, 17.2)

<u>*Accreditation system*</u>

10. System that has its own rules of procedure and management for carrying out accreditation.

11. Note – Accreditation of certification bodies is normally awarded following successful assessment and is followed by appropriate surveillance.
(Based on ISO Guide 2, paragraph 17.1)

<u>*Arrangement*</u>

12. A cooperative mechanism established by two or more parties be they governmental, private or non-governmental entities.

<u>*Audit*</u>

13. A systematic and functionally independent examination to determine whether activities and related results comply with planned objectives.
(Based on Codex Alimentarius, Principles for Food Import and Export Certification and Inspection, CAC/GL 20)

<u>*Certification*</u>

14. Procedure by which a third party gives written or equivalent assurance that a product, process or service conforms to specified requirements. Certification may be, as appropriate, based on a range of inspection activities which may include continuous inspection in the production chain.
(Based on ISO Guide 2, 15.1.2 and Principles for Food Import and Export Certification and Inspection, CAC/GL 20)

<u>*Certification body*</u>

15. Competent and recognized body that conducts certification. A certification body may oversee certification activities carried out on its behalf by other bodies.
(Based on ISO Guide 2, 15.2)

<u>*Chain of custody*</u>

16. The set of measures which is designed to guarantee that the product put on the market and bearing the ecolabel logo is really a product coming from the certified fishery concerned. These measures should thus cover both the tracking/traceability of the product all along the processing, distribution and marketing chain, as well as the proper tracking of the documentation (and control of the quantity concerned).

<u>*Complaint*</u>

17. *An objection by a person or body to a decision regarding accreditation, de-accreditation, certification or de-certification.*

<u>*Conformity assessment*</u>

18. *Any activity concerned with determining directly or indirectly that relevant requirements are fulfilled.*
19. *Notes: Typical examples of conformity assessment activities are sampling, testing and inspection; evaluation, verification and assurance of conformity (supplier's declaration, certification); registration, accreditation and approval as well as their combinations.*
    (ISO Guide 2, 12.2)

<u>*Decision*</u>

20. *Any resolution by an accreditation or certifying body or arrangement concerning the rights and obligations of a person or body.*

<u>*Ecolabelling scheme*</u>

21. *Ecolabelling schemes entitle a fishery product to bear a distinctive logo or statement which certifies that the fish has been harvested in compliance with conservation and sustainability standards. The logo or statement is intended to make provision for informed decisions of purchasers whose choice can be relied upon to promote and stimulate the sustainable use of fishery resources.*

<u>*Standard for certification*</u>

22. *Document approved by a recognized organization or arrangement, that provides, for common and repeated use, rules, guidelines or characteristics for products or related processes and production methods, with which compliance is not mandatory under international trade rules. It may also include or deal exclusively with terminology, symbols, packaging, marking or labelling requirements as they apply to a product, process or production method.*
    (Based on TBT agreement, Annex 1, para.2)

    In these guidelines, unless otherwise qualified, the word standard refers to a standard for certification. The standard for certification will include requirements, criteria and performance elements in a hierarchical arrangement. For each requirement, one or more substantive criteria should be defined. For each criterion, one or more performance elements should be provided for use in assessment.

<u>*Standard-setting organization or arrangement*</u>

23. *Organization or arrangement that has recognized activities in standard setting.*
    (Based on ISO Guide 2, paragraph 4.3)

<u>*Third party*</u>

24. *Person or body that is recognized as being independent of the parties involved, as concerns the issue in question.*
    (ISO/IEC Guide 2:1996)

Unit of certification

25. *The 'unit of certification' is the fishery for which ecolabelling certification is sought, as specified by the stakeholders who are seeking certification. The certification could encompass: the whole fishery, where a fishery refers to the activity of one particular gear-type or method leading to the harvest of one or more species; a sub-component of a fishery, for example a national fleet fishing a shared stock; or several fisheries operating on the same resources. The 'stock under consideration' exploited by this fishery (unit of certification) may be one or more biological stocks as specified by the stakeholders for certification. The certification applies only to products derived from the 'stock under consideration' (see Para. 30). In assessing compliance with certification standards, the impacts on the 'stock under consideration' of all the fisheries utilizing that 'stock under consideration' over its entire area of distribution are to be considered.*

7.4.5 Minimum substantive requirements and criteria for ecolabels

This section of the guidelines sets out the minimum substantive requirements and criteria for assessing whether a fishery can be certified and awarded an ecolabel. It keeps open the option for ecolabelling schemes to apply additional or more stringent requirements and criteria.

The drafting of this section was informed by the Code of Conduct for Responsible Fisheries, the UN Fish Stocks Agreement, the principles and criteria of the MSC as well as those elaborated by the Nordic Technical Working Group on Ecolabelling Criteria (2000), a group set up by the Nordic Council of Ministers in 2000. There were also several expert consultations (see References) that contributed to the finely elaborated text of this core section of the guidelines on the definition and assessment of a sustainable fishery. Minimum requirements are specified for each of three areas: the management systems, the fishery and associated 'stock under consideration', and ecosystem considerations. This is in keeping with the idea that both the process and the outcome of management need to be considered. The requirements and criteria exclude economic, social or safety-at-sea considerations.

This section acknowledges that conventional stock assessment methods may not be possible nor necessarily appropriate in all cases and that 'less elaborate' methods may be used (paragraph 32a). However, attention is also drawn to the need to consider the amount of uncertainty in the final outcome of the assessment and to apply the precautionary approach accordingly. The section explicitly recognises the value of traditional knowledge provided its validity can be objectively verified.

There was considerable concern amongst some countries, especially some developing countries, about the inclusion of 'Ecosystem considerations' in the minimum requirements. This arose from the knowledge that, in many countries, current knowledge on ecosystems and ecosystem impacts is weak because of the lack of data and research due to financial and human resources constraints. The inclusion of ecosystem considerations could therefore become an effective barrier to obtaining an ecolabel and consequently a barrier to trade. This section, within the core of the guidelines, therefore represents a reasonable compromise between the position of some countries seeking more stringent requirements and criteria and others that wished to see ecosystem considerations entirely omitted from the guidelines.

In reference to the modifications to this section, the inland fisheries expert group noted that enhanced fisheries may involve a number of techniques, some of which are permanent or nearly so, e.g. species introductions and habitat modification, and some of which could be temporary. The sustainability of the target species, therefore, could depend on the maintenance of the enhancements. In the special case of culture-based fisheries, where the fishery is solely maintained by stocking from aquaculture facilities, the experts concluded that sustainability of the target species would not be the focus of an ecolabelling programme. Instead, sustainability would relate primarily to assuring optimal production in the natural ecosystem and management in a manner to conserve biodiversity and ecosystem functions (FAO, 2006).

Introduction

26. *The following sets forth the minimum substantive requirements and criteria for assessing whether a fishery can be certified and an ecolabel awarded to a fishery. Ecolabelling schemes may apply additional or more stringent requirements and criteria related to sustainable use of the resources. The requirements and criteria presented below are to be based on and interpreted in accordance with the current suite of agreed international instruments in particular the 1982 UN Convention on the Law of the Sea, the 1995 UN Fish Stocks Agreement and the 1995 Code of Conduct for Responsible Fisheries, as well as related documentation including the 2001 Reykjavik Declaration on Responsible Fisheries in the Marine Ecosystem.*
27. *Requirements are specified for each of three areas: the management systems, the fishery and associated 'stock under consideration' for which certification is being sought, and consideration of serious impacts of the fishery on the ecosystem. Criteria and related measurable performance indicators and a corresponding monitoring system should be established in order to assess the conformity of the fishery concerned with the requirements and the criteria of the ecolabelling scheme. In developing and applying the criteria and assessing the conformity of the fishery with the standard of certification, the views and opinions of States, RFMOs and FAO should be fully considered.*

Management systems

28. *Requirement: The fishery is conducted under a management system which is based upon good practice and that ensures the satisfaction of the requirements and criteria described in Paragraph 29. The management system and the fishery operate in compliance with the requirements of local, national and international laws and regulations, including the requirements of any regional fisheries management organization that manages the fisheries on the 'stock under consideration'.*
    - *28.1 For the 'stock under consideration' there are documented management approaches with a well based expectation that management will be successful taking into account uncertainty and imprecision.*
    - *28.2 There are objectives, and as necessary, management measures to address pertinent aspects of the ecosystem effects of fishing as per paragraph 31.*
29. *The following criteria will apply to management systems for any fisheries, but it must be recognized that special consideration needs to be given to small-scale fisheries with respect to the availability of data and with respect to the fact that management systems can differ substantially for different types and scales of fisheries (e.g. small scale through to large scale commercial fisheries).*
    - *29.1 Adequate and reliable data and/or information are collected, maintained and assessed in accordance with applicable international standards and practices for evaluation of the current state and trends of the stocks (see below: Methodological aspects). This can include relevant traditional, fisher or community knowledge, provided its validity can be objectively verified.*
    - *29.2 In determining suitable conservation and management measures, the best scientific evidence available is taken into account by the designated authority, as well as consideration of relevant traditional fisher or community knowledge, provided its validity can be objectively verified, in order to evaluate the current state of the 'stock under consideration' in relation to, where appropriate, stock specific target and limit reference points.*
    - *29.2bis: Taking due account of paragraph 32, for the 'stock under consideration' the determination of suitable conservation and management measures should include or take account of:*
        - *Total fishing mortality from all sources is considered in assessing the state of the 'stock under consideration', including discards, unobserved mortality, incidental mortality, unreported catches and catches in other fisheries.*
        - *Management targets are consistent with achieving MSY (or*

a suitable proxy) on average, or a lesser fishing mortality if that is optimal in the circumstances of the fishery (e.g. multi-species fisheries) or to avoid severe adverse impacts on dependent predators.

- *The management system should specify limits or directions in key performance indicators (see 30.2), consistent with avoiding recruitment overfishing or other impacts that are likely to be irreversible or very slowly reversible, and specify the actions to be taken if the limits are approached or the desired directions are not achieved.*

29.3 Similarly, data and information, including relevant traditional fisher or community knowledge, provided its validity can be objectively verified, are used to identify adverse impacts of the fishery on the ecosystem, and timely scientific advice is provided on the likelihood and magnitude of identified impacts (see Paragraph 31).

29.4 The designated authorities adopt and effectively implement appropriate measures for the conservation and sustainable use of the 'stock under consideration' based on the data, information, and scientific advice referred to in the preceding bullets. Short-term considerations should not compromise the long-term conservation and sustainable use of fisheries resources.

29.5 An effective legal and administrative framework at the local, national or regional level, as appropriate, is established for the fishery[8] and compliance is ensured through suitable mechanisms for monitoring, surveillance, control and enforcement (see also Paragraph 6).

29.6 In accordance with the Code of Conduct Article 7.5, the precautionary approach is being implemented to protect the 'stock under consideration' and the aquatic environment. Inter alia this will require that the absence of adequate scientific information should not be used as a reason for postponing or failing to take conservation and management measures. Further, relevant uncertainties are being taken into account through a suitable method of risk assessment. Appropriate reference points are determined and remedial actions to be taken if reference points are approached or exceeded are specified.

'Stocks under consideration'

30. Requirement: The 'stock under consideration' is not overfished, and is maintained at a level which promotes the objective of optimal utilization and maintains its availability for present and future generations, taking into account that longer term changes in productivity can occur due to natural variability and/or impacts other than fishing. In the event that biomass drops well below such target levels, management measures (Code of Conduct Article 7.6) should allow for restoration within reasonable time frames of the stocks to such levels (see also paragraph 29.2 bis).

The following criteria are applicable:

30.1 The 'stock under consideration' is not overfished if it is above the associated limit reference point (or its proxy).

30.2 If fishing mortality (or its proxy) is above the associated limit reference point, actions should be taken to decrease the fishing mortality (or its proxy) below that limit reference point.

30.3 The structure and composition of the 'stock under consideration' which contribute to its resilience are taken into account.

30.4 In the absence of specific information on the 'stock under consideration', generic evidence based on similar stocks can be used for fisheries with low risk to that 'stock under consideration'. However, the greater the risk the more specific evidence is necessary to ascertain the sustainability of intensive fisheries.

Ecosystem considerations

31. Requirement: Adverse impacts of the fishery on the ecosystem should be appropriately assessed and effectively addressed. Much greater scientific uncertainty is to be expected in assessing possible adverse ecosystem impacts of fisheries. This issue can be addressed by taking a 'risk assessment/risk management approach'. For the purpose of development of ecolabelling schemes, the most probable adverse impacts should be considered, taking into account available scientific information, and traditional, fisher or community knowledge provided that its validity can be objectively verified. Those impacts that are likely to have serious consequences should be addressed. This may take the form of an immediate management response or further analysis of the identified risk. In this context, full recognition should be given to the special circumstances and requirements in developing countries and countries in transition, including financial and technical assistance, technology transfer, and training and scientific cooperation.

The following criteria are to be interpreted in the context of avoiding high risk of severe adverse impacts.

31.1 Non target catches, including discards, of stocks other than the 'stock under consideration' are monitored and should not threaten these non-target stocks with serious risk of extinction; if serious risks of extinction arise, effective remedial action should be taken.

31.2 The role of the 'stock under consideration' in the food-web is considered, and if it is a key prey species in the ecosystem, management measures are in place to avoid severe adverse impacts on dependent predators.

31.3 There is knowledge of the essential habitats for the 'stock under consideration' and potential fishery impacts on them. Impacts on essential habitats and on habitats that are highly vulnerable to damage by the fishing gear involved are avoided, minimised or mitigated (Code of Conduct 7.2.2). In assessing fishery impacts, the full spatial range of the relevant habitat should be considered, not just that part of the spatial range that is potentially affected by fishing.

31.4 In the absence of specific information on the ecosystem impacts of fishing for the unit of certification, generic evidence based on similar fishery situations can be used for fisheries with low risk of severe adverse impact. However, the greater the risk the more specific evidence is necessary to ascertain the adequacy of mitigation measures.

Methodological aspects

Assessing current state and trends in target stocks

32. There are many ways in which state and trends in stocks may be evaluated, that fall short of the highly quantitative and data-demanding approaches to fish stock assessment that are often used in developed countries. However it should be noted that, to the extent that the application of such methods may result in greater uncertainty about the state of the 'stock under consideration', more precautionary approaches to managing fisheries on such resources could be required which may necessitate lower levels of utilization of the resource. There is a variety of management measures commonly used in small scale or low value fisheries that nonetheless can achieve quite adequate levels of protection for stocks in the face of uncertainty about the state of the resource. A past record of good management performance could be considered as supporting evidence of the adequacy of the management measures and the management system.

7.4.6 Procedural and institutional aspects

This part of the guidelines addresses the three principal procedural and institutional matters that any ecolabelling scheme should encompass: (1) the setting of certification standards, (2) the accreditation of independent certifying bodies, and (3) the certification that a fishery and the product chain of custody are in conformity with the required standard and procedures.

Except for the issue of an independent panel as ultimate appellate body, consensus on this part of the guidelines was reached early in the negotiation process. From the beginning of the ecolabelling discussion in FAO fora, countries supported the principle of independent and transparent third party certification through competent, reliable and accountable bodies. Many of the provisions in this section are geared towards assuring the application of this principle.

In this context, it is notable that some countries felt strongly the need for MSC to adjust its governance structure in order to assure complete independence between its functions as the owner and promoter of an ecolabelling scheme and the functions of accreditation and certification, including the sensitive aspect of dealing with complaints. In order to achieve consistency with the FAO guidelines, MSC has subsequently appointed an independent objections panel chair to ensure the impartiality of any panels formed to hear appeals against proposed certification decisions. This appointment served to separate the objections process and related decisions from the MSC's Board of Trustees (MSC, 2006).

Further, there was broad consensus on the need to engage all interested parties in the standard-setting process in a consultative and participatory manner. A number of governments and industry groups felt that the MSC process did not accomplish this requirement in its initial phase.

Options for governance structures

The guidelines are not overly prescriptive on other aspects of the governance structure beyond the above-noted separation between ownership and accreditation functions. This allows for ecolabelling schemes to be established by a government, an intergovernmental organisation, a non-governmental organisation, or a private industry association. There are also various options for the geographical range of a scheme – national, regional or international in scope (paragraph 37).

Guidelines for the setting of standards for sustainable fisheries

The setting of standards is among the most critical tasks of any ecolabelling scheme. The standards reflect the objectives for sustainable fisheries that are being pursued through the scheme. Standards comprise quantitative and qualitative indicators of the governance system or management regime of a fishery as well as of its outcome in terms of sustainable fisheries and conservation of marine fishery resources and related ecosystems (paragraph 40).

The principal normative basis for the procedural requirements in standard setting is given by the WTO TBT, ANNEX 3 *Code of Good Practice for the Preparation, Adoption and Application of Standards* and the ISO/IEC Guide 59 Code of good practice for standardisation of 1994. More recent work in this area has been done by the ISEAL Alliance which published in early 2006 the final version of a Code of Good Practice for Setting Social and Environmental Standards (isealalliance.org). At the core of standard-setting norms are the ideas of consultation and participation of interested parties in a transparent and well-informed process of standard setting that provides for appropriate notification and minimum time periods for commenting.

The functions of a standard-setting organisation or arrangement include the setting, reviewing, revising, assessing, verifying and approving of standards. Where there is no specialised body, the organisational structure of a standard-setting arrangement should include, *inter alia*, a technical committee of independent experts and a consultation forum whose mandates are well established (paragraphs 44 and 45).

The guidelines explicitly identify the various interested parties that ideally should participate in the development of standards of sustainable fisheries. These include representatives of fisheries management authorities, the fishing industry, fishers organisations, the scientific community, environmental interest groups, fish processors, traders and retailers as well as consumer associations (paragraph 54). The inland fisheries expert group added fishing communities and hatchery managers to this list of interested parties.

An innovative feature is the requirement that, in developing or revising a standard, an appropriate procedure should be put in place to validate the standard with respect to the minimum requirements for sustainable fisheries as laid out in the guidelines. There is also a call for standards to not encompass criteria or requirements that are of no relevance for sustainable fisheries or could cause unnecessary barriers to trade or mislead the consumer (paragraph 63).

Guidelines for accreditation

The purpose of accreditation is to provide assurance that certification bodies responsible for conducting conformity assessments with sustainability standards and chain of custody requirements are competent to carry out such tasks. The guidelines lay down the requirements for accreditation organisations to perform this task professionally in a transparent, impartial, independent, and accountable fashion. The primary normative basis is the ISO Guide 61, *General Requirements for assessment and accreditation of certification/registration bodies*, 1996.

The conditions for maintaining, extending, suspending and withdrawal of accreditation are also spelled out in the guidelines as is the responsibility of the accreditation body in relation to the use of accreditation marks, symbols and logos and how to prevent their misleading use in advertisements, etc.

Past experience with ecolabelling schemes points to the importance of having solid procedures to address and resolve complaints in an impartial and independent manner. In this regard, the guidelines spell out the need for the establishment of an impartial and independent committee which, in the first instance, should attempt to resolve any complaints through discussion or conciliation and, if this fails, in the second instance provide a written ruling to the accreditation body and the parties concerned (paragraph 83). The guidelines, however, explicitly state that this provision would not exclude recourse to other forms of legal and administrative processes as provided for in national legislation or international law (paragraph 86).

Guidelines for certification

Certification is an integral and indispensable part of any ecolabelling scheme. In respect to fisheries ecolabelling schemes, it provides assurance to buyers and consumers that a certain fish or fishery product comes from a fishery that conforms to the established standard for a sustainable fishery. In keeping with the Principles, impartial certification based on an objective assessment of all relevant factors ensures that ecolabels convey truthful information. This is a necessary condition for the ecolabelling scheme to attain its objectives.

The guidelines provide for two types of certification, certification of the fishery itself, including the production of stocking material in the case of inland fisheries, and certification of the chain of custody between the time the fish is harvested and the time the fish or fishery product is sold to the final consumer. The chain of custody assessment examines whether adequate measures are in place to identify fish from a certified fishery at subsequent stages of fish processing, distribution and marketing. While separate certificates may be issued for the fishery and for the chain of custody, fish and fishery products that are labelled to indicate to the consumer their origin from a sustainable fishery require both types of assessments.

As is the case for accreditation organisations, the guidelines lay down the requirements for certification bodies to perform their tasks professionally in a transparent, impartial, independent, and accountable fashion. The primary normative basis includes ISO Guide 62, *General Requirements for bodies operating assessment and certification/registration of quality systems*, 1996, ISO/IEC Guide 65, *General requirements for bodies operating product certification systems*, 1996, and Article 5 of the WTO Agreement on Technical Barriers to Trade.

Beyond the general ISO requirements appropriately adapted to the case of sustainable fisheries, the guidelines contain specific provisions that acknowledge the great diversity of situations and conditions under which fisheries are conducted and managed. To ensure non-discrimination, the access to the services of a certification body should be open to all types of fisheries, whether managed by a regional, governmental, parastatal or non-governmental fisheries management organisation or arrangement. Further, access to certification should not be conditional upon the size or scale of the fishery, nor should certification be conditional upon the number of fisheries already certified (paragraph 112).

Non-discrimination in access to certification services is also the intent of the provision on the certification fee structure (paragraph 125).

> In establishing the fee structure and in determining the specific fee of a certification assessment, the certification body should take into account, inter alia, the requirements for accurate and truthful assessments, the scale, size and complexity of the fishery or chain of custody, the requirement of non-discrimination of any client, and the special circumstances and requirements of developing countries and countries in transition.

Given the highly dynamic nature of fisheries, the guidelines contain detailed provisions on the maintenance, renewal and possible suspension and withdrawal of certification. They call for periodic surveillance and monitoring of the fishery and chain of custody at appropriate time intervals (paragraph 128), prompt notification by the client of intended changes to the management of the fishery or chain of custody (paragraph 129), and reassessments in the event of changes significantly affecting the status and management of the fishery or chain of custody, or if analysis of complaints and other information indicates that the certified fishery and/or chain of custody no longer comply with the required standard (paragraph 130).

The period of validity of a certificate should not exceed five years in the case of a fishery and three years in the case of the chain of custody (paragraph 131). Given regular monitoring and auditing exercises and a full reassessment, the validity of certification can be renewed for the same time periods (paragraph 132).

The guidelines place the responsibility on the certifying body to specify the conditions under which certification may be suspended or withdrawn (paragraph 133). If a certification is withdrawn or suspended, the certifying body should require that a certified fishery and/or chain of custody discontinues use of all advertising matter that contains any reference thereto and returns any certification documents. The certification body also has the responsibility of informing the public about the withdrawal or suspension after the appeals process is exhausted (paragraph 134).

Assurance of the chain of custody is complex in fisheries because of the often large number of fishing vessels, landing places and localities of processing, marketing and distribution. In recognition of rapid technological progress in traceability, the physical segregation of certified from non-certified fish and fishery products was not considered to be an indispensable requirement in all instances, as had initially been proposed by the expert consultation. However, the guidelines provide for detailed chain of custody requirements and monitoring and auditing procedures by the certification body (paragraphs 135–140).

In recognition of the proliferation of unsubstantiated product claims and logos in respect to fish and fishery products, the guidelines are very specific about the responsibilities of the certification body, accreditation body or owner of the ecolabelling scheme over the use and control of certification claim, symbol and logo. They are required to ensure that symbols or logos should not relate to claims that are of no relevance for sustainable fisheries or could cause barriers of trade or mislead the consumer (paragraph 141). Only products from certified sources can carry a mark/claim/logo (paragraph 142), no fraudulent or misleading use can be made with their use and display (paragraph 143), and suitable action must be taken to deal with incorrect references to the certification system or misleading use of symbols and logos found in advertisements, catalogues, etc. (paragraph 145).

Resolution of complaints and appeals

Within the procedural part of the guidelines, this section on the resolution of complaints and appeals relating to certification has been intensely discussed in the negotiation process. The Expert Consultation proposed to include in this section an independent panel external to the ecolabelling scheme as an ultimate appellate body. This panel which would consider in last instance appeals of a finding or decision concerning the conformity assessment only, thus excluding the chain of custody assessment, would have been convened and serviced by FAO. All costs relating to it would have had to be borne by the appellant.

The idea of the independent panel was derived as an added precaution to ensure independence in addressing complaints, particularly given the high proportion of fish and fishery products internationally traded, the likelihood of cross-border complaints, and the often paramount role of governments in fisheries management. Governments, as a general rule, do not like to be assessed by, and subject to, decision-making through non-governmental entities, especially on a matter as complex and controversial as fisheries management.

There was not unanimous support for the inclusion of an independent panel in the FAO guidelines. Several countries felt that possible recourse to other forms of legal and administrative processes as provided for in national legislation or international law would provide adequate safeguards to seek redress in the case of flawed rulings within the ecolabelling scheme's internal complaint and appeal procedures. Other countries expressed a strong desire for an independent panel, probably because of concerns about access to other systems of ruling, e.g. the courts, and timely rulings.

However, after careful examination and review of past practices, it became evident that FAO's envisaged role in servicing such an independent panel would be in conflict with the Organization's basic text. While FAO's basic text foresees the convening of expert panels, their constitutional purpose, as evidenced also by past practice, is to provide advice to the Director General on specific subjects. The independent panel, on the other hand, is an appellate body whose purpose is of a judicial nature and not to give advice to the Director General.

7.5 Conclusions

In the area of ecolabelling, the FAO guidelines for marine capture fisheries are a unique voluntary international instrument that establishes minimum standards in procedural and substantive terms. The guidelines can help to prevent the proliferation of non-credible ecolabels, contribute to the creation of an equal playing field by recognising the special conditions and requirements of fisheries in developing countries and countries in transition, provide clarity on equivalence of ecolabelling schemes and non-discrimination, avoid unnecessary barriers to trade, and establish the legitimacy of ecolabelling applied to fisheries.

Time will show whether the guidelines will succeed in all these aspects. One area of special attention for FAO will be the promotion of participation of developing country small-scale fisheries in ecolabelling schemes. These fisheries support millions of fishers and contribute directly and significantly to poverty alleviation and food security. The sustainability of these fisheries is critical for maintaining and enhancing the contribution of fisheries to national well-being.

7.6 References

COCHRANE, K., WILLMANN, R. (2000) Eco-labelling in fisheries management. In: *Current Fisheries Issues and the Food and Agriculture Organization of the United Nations* (eds M.H. Nordquist and J.N. Moore). Martinus Nijhoff Publishers, The Hague/Boston, MA/ London, 583–615.

FAO (2000) *The State of the World Fisheries and Aquaculture*. FAO, Rome.

FAO (2003) Report of the Expert Consultation on the Development of International Guidelines for Ecolabelling of Fish and Fishery Products from Marine Capture Fisheries. Rome, Italy, 14–17 October 2003. *FAO Fisheries Report. No. 726*. FAO, Rome.

FAO (2005a) Report of the Technical Consultation on International Guidelines for the Ecolabelling of Fish and Fishery Products from Marine Capture Fisheries. Rome, 19–22 October 2004. Rapport de la Consultation technique sur les directives internationales pour l'étiquetage écologique du poisson et des produits des pêches de capture marines. Rome, 19–22 octobre 2004. Informe de la Consulta Técnica sobre las Directrices Internacionales para el Ecoetiquetado de Pescadoy Productos Pesqueros de la Pesca de Captura Marina. Roma, 19–22 de octubre de 2004. *FAO Fisheries Report/FAO Rapport sur les pêches/ FAO Informe de Pesca*. No. 760. FAO, Rome/Roma.

FAO (2005b) Report of the twenty-sixth session of the Committee on Fisheries. Rome, 7–11 March 2005. *FAO Fisheries Report. No. 780*. FAO, Rome.

FAO (2006) Report of the Expert Consultation on the Development of International Guidelines for the Ecolabelling of Fish and Fishery Products from Inland Capture Fisheries. Rome, 23–26 May 2006. *FAO Fisheries Report*. No. 804. FAO, Rome.

FAO (2008a) *The State of the World Fisheries and Aquaculture*. FAO, Rome.

FAO (2008b) Report of the Expert Consultation on the FAO Guidelines for Ecolabelling. Rome, 3–4 March 2008. *FAO Fisheries Report. No. 864*. FAO, Rome.

ICSF (1998) *Fish stake – the pros and cons of the Marine Stewardship Council initiative: a debate from the pages of SAMUDRA Report*. International Collective in Support of Fishworkers. Chennai, India, 6.

MSC (2006) *Fish4Thought*, July 2006. Marine Science Council, London.

Nordic Technical Working Group on Fisheries Ecolabelling Criteria (2000) *An Arrangement for the Voluntary Certification of Products of Sustainable Fishing. Final Report*. Copenhagen, 21 June 2000.

WARD, T. J., PHILLIPS, B. (eds) (2008) *Seafood Ecolabelling: Principles and Practice*. Wiley-Blackwell, Chichester/Ames, IA.

WESSELLS, C.R., COCHRANE, K., DEERE, C., WALLIS, P., WILLMANN, R. (2001) Product certification and ecolabelling for fisheries sustainability. *FAO Fisheries Technical Paper. No. 422*. FAO, Rome.

WORLD BANK AND FAO (2009) *The sunken billions – the economic justification for fisheries reform*. World Bank. Washington D.C.

WTO (2003). Committee on Trade and Environment. Report to the 5th Session of the WTO Ministerial Conference in Cancún. Paragraphs 32 and 33 of the Doha Ministerial Declaration (WT/CTE/8 11 July 2003). Geneva, WT.

8

Voluntary environmental and social labels in the food sector

Pascal Liu, Food and Agriculture Organization of United Nations (FAO), Italy

Abstract: Since the late 1990s environmental and social labels have become widespread in developed countries. The most common environmental and social certification labels in the food sector are organic agriculture, fair-trade and the Rainforest Alliance. Sales of labelled products have risen rapidly over the past ten years and the outlook is for continued market growth, albeit at a lower rate. Producers may benefit from better market access and, in many cases, a price premium. Additional benefits can include higher profitability as well as non-economic benefits such as enhanced self-esteem, social image and relationships with business partners. However, a substantial investment in time and money is often needed for certification and compliance with standards. The cases of organic bananas and fair-trade coffee illustrate the potential benefits and costs of these labelling systems for farmers in developing countries.

Key words: certification, labels, markets, fair trade, organic.

8.1 Introduction

Since the early 1990s, a variety of voluntary certification labels have become available to the agricultural industry. These include the labels associated with voluntary environmental and social standards. Such standards have been developed by a wide array of organizations, from both the public and private sectors, at local, national or international level (FAO 2007). Private sector labels include labels developed by businesses (e.g. food manufacturers and retailers) and those created by not for profit NGOs. The latter type of labels covers a wide range of issues such as environmental protection, labour rights, safety and health at work, social equity and the welfare of local communities. A growing number of agricultural producers and traders have sought to obtain certification against one or more of these standards for a variety of reasons. Some of the labelling schemes may generate a price premium. Other possible benefits lie in improved market access and stability. Some schemes help rationalize production, reduce costs, improve labour management and enhance the morale and participation of workers. Others help preserve productive natural resources. Sometimes the main reason for adoption is the need to improve the company's image and show its commitment to social responsibility. Among the environmental and social labels, the most common in the agriculture sector are the Rainforest Alliance, organic agriculture and fair trade labels.

This chapter only deals with labels that are associated with voluntary environmental and social certification programmes,[1] which producers are free to adopt or not. Labels related to mandatory governmental standards (officially named 'technical regulations') are outside its scope.

8.2 Background: environmental and social issues in agriculture

The rise of certification and labelling schemes aiming at sustainable agriculture results to a large extent from growing consumer awareness of the adverse environmental and social effects of large-scale commercial farming. In particular, the expansion and intensification of production in large plantations in the 1980s and early 1990s gave rise to a series of environmental problems. The expansion was sometimes done at the expense of forest or other natural vegetation. More importantly, agricultural production for export is generally intensive, with high levels of external inputs, and often takes place in monoculture plantations organized along agro industrial lines. Most plantations rely on the frequent use of agrochemicals to maintain fertility and limit losses caused by pests. Inappropriate production practices have often led to pollution of land, watercourses and aquifers, and a reduction in biological diversity.

As consumers have become increasingly sensitive to environmental issues, intensive agricultural production has attracted growing attention. Some industries (e.g. bananas, palm oil) have come under close scrutiny. Strong pressure from non-governmental organizations (NGO), negative media coverage and a shift in consumer preference towards 'ecofriendly' products have led some companies to take measures to reduce the adverse impacts on the environment. Solutions have been sought to the most pressing problems. The management of input and output flows has been rationalized in many farms. Waste disposal has improved considerably since the late 1990s. Collection of plastics, composting of organic rejects and filtering of wastewater have become common practices on many plantations.

---

[1]In certification, compliance with the standard is verified by an independent third party.

However, the pollution caused by the intensive use of agrochemicals in monoculture production remains a challenge, as changes in input use may directly affect productivity. Monoculture attracts a wide range of pests and diseases, notably fungal diseases, which are difficult to combat in tropical climates.

Beside its negative environmental impact, the use of pesticides may also have adverse effects on the health of plantation workers and neighbouring communities. Further, the long-term toxicity of an authorized pesticide may be discovered only many years after its approval was granted. Various cases of soil contamination by the indiscriminate use of pesticides that were legal for long periods have been reported in a number of countries.

The agriculture sector has faced social problems related to the non-respect of labour rights. In several instances, the conventions of the International Labour Organization (ILO) and even national labour laws were not enforced, leading to abuses such as child work, excessive working hours, discrimination, sexual harassment, non-respect of health and safety regulations and absence of provision of medical assurance. Another frequently debated social issue in agriculture is the right to freedom of association and collective bargaining, as formulated in ILO conventions No. 87 (1948) and No. 98 (1949). These problems have coincided with growing consumer awareness of the 'ethics' of food production and trade due in part to the sensitization campaigns launched by various NGOs working in areas such as human rights, social development and 'fair trade'. Issues such as conditions of work, wages of farm labour or the price paid to small producers in developing countries attracted public attention in developed countries. Consumer associations and other groups now want guarantees that workers' health is not put at risk by the lack of adequate safety measures on the farm or the use of pesticides known to be hazardous. They are increasingly interested in labour rights issues such as freedom of association or the right to join an independent trade union, as well as in 'fair' remuneration of farm workers and small producers.

Under pressure from NGO campaigns, retailer demands and increased consumer awareness of ethical trade in the importing countries, companies have taken steps to improve the situation of their work force. This tendency was first apparent in the marketing of imported handicraft products, as exemplified by shops guaranteeing their customers that their rugs were not produced using child or forced labour. In the 1990s, the movement reached larger manufacturers of consumer goods (e.g. garments and sport shoes), demanding that they exert a closer monitoring of the working conditions in their subsidiaries worldwide. Social concerns have also reached the agricultural sector and some progress has been observed in recent years.

8.3 Main environmental and social labelling schemes in agriculture

8.3.1 Basic principles of certification and labelling schemes

Certification is a written guarantee by an independent certification body that a production process or a product meets the criteria or requirements contained in a certain standard. The certification body is a third party that has no interest in the economic relationship between the supplier and buyer. The basic elements of a certification system (also called certification 'programme' or 'scheme') are the standard and the system to control the compliance of the certified entity with the standard. The object of certification can be a product or a process. Environmental and social labels are generally aimed at the production process (and sometimes also the trading process, as in fair trade standards). These standards may focus on environmental issues such as soil conservation, water protection, pesticide use or waste management; on social issues such as worker rights, occupational health and safety; or on other issues such as food safety. The improvements can result in the protection of local resources, healthier workers and other benefits for producers, consumers and local communities. The certification is voluntary when producers freely decide whether or not they want to certify their production process and facility.

Certification is used to demonstrate that the product has been produced in accordance with a certain process or has certain characteristics. It can differentiate the product from other products, which can be helpful to promote the product in different markets, improve its market access and, in some cases, fetch a better price. Certification is mainly used when the producer and the consumer are not in direct contact, as in the international market. In those cases where there are doubts on the effectiveness of the regulatory system of the exporting country, certification may help exporters create trust (Cuffaro and Liu, 2007).

Producers can choose among many different types of certification. The decisions on whether or not to seek certification and what type of certification to choose are important choices that influence farm management, investments and marketing strategies. Each certification programme has different objectives and thus different requirements that the producer must comply with in order to be certified. The cost of complying with the standard and of certification depends on the types of changes the producer will have to make and on the type of certification programme chosen. In general, the cost of certification is based on the time spent by the inspector(s) doing the farm inspection (farm audit) and on their travel expenses.

8.3.2 Certification labels frequently used in the agricultural export industry

There are a number of certification and labelling programmes that apply to agricultural exports. This chapter covers those environmental and/or social labels which are the most significant to agricultural trade in terms of certified quantities, namely organic agriculture, fair-trade and Rainforest Alliance. The selected labelling schemes are all voluntary, i.e. producers and traders are free to choose whether to seek certification or not. However, these schemes differ widely in terms of ownership, objectives, scope, requirements, criteria, indicators and monitoring procedures. All are privately owned standards, except for organic agriculture. The Rainforest Alliance standard is a single standard, owned by a non-governmental

Table 8.1 Main characteristics of selected labelling schemes

	Organic	Fair trade	Rainforest Alliance
Number of standards:	> 10	> 4	1
Ownership:	Governments, NGOs	NGOs, Certification bodies	RA-SAN (NGO)*
Focus of standard:	Environmental	Social equity	Environmental
Countries where it applies:	All	Developing countries	Developing countries
Certification by:	Accredited certification bodies (CB)	FLO-Cert for FLO** standard.CB for their own standards	Sustainable Farm Certification, Intl
Main beneficiaries:	All types of farms	Small farmers	Large farms

*Rainforest Alliance-Sustainable Agriculture Network
**Fairtrade Labelling Organizations International

organization, while there are several fair trade standards. Similarly, there is a wide array of organic agriculture standards, some privately owned, some governmental, some intergovernmental (Table 8.1).

Organic agriculture is a production method which manages the farm and its environment as a single system. It utilizes both traditional and scientific knowledge to enhance the health of the agro-ecosystem in which the farm operates. Organic farms rely on the use of local natural resources and the management of the ecosystem rather than external agricultural inputs such as mineral fertilizers and agrochemicals. Organic agriculture therefore rejects synthetic chemicals and genetically modified inputs. It promotes sustainable traditional farming practices that maintain soil fertility such as fallow and nutrient recycling (e.g. compost and crop litter).

There is a variety of organic agriculture standards (see Chapter 6). Historically, the first standards were developed by non-governmental organizations (e.g. organic farmer associations, trade associations, certification bodies). Then, as the market for organics grew, governments started to regulate organic labelling and develop national standards. France was among the first governments to adopt a regulation on organic farming. Now, most developed countries have a public standard and regulations governing the production, marketing and labelling of organic products. Finally, some intergovernmental entities have adopted laws and standards. The European Union adopted it in 1991 (Regulation EEC 2092/91). In 1999, the Committee on Food Labelling of the Codex Alimentarius Commission adopted Guidelines for the Production, Processing, Labelling and Marketing of Organically Produced Foods. According to the Codex definition:

> *organic agriculture is a holistic production management system which promotes and enhances agro-ecosystem health, including biodiversity, biological cycles, and soil biological activity. It emphasises the use of management practices in preference to the use of off-farm inputs, taking*

into account that regional conditions require locally adapted systems. This is accomplished by using, where possible, agronomic, biological, and mechanical methods, as opposed to using synthetic materials, to fulfil any specific function within the system. (CAC, 1999)

Organic agriculture is one of several approaches to sustainable agriculture, and many of the techniques used (e.g. inter-cropping, rotation of crops, double-digging, mulching, integration of crops and livestock) are practised under various agricultural systems. What makes organic agriculture unique, as regulated under various laws and certification programmes, is that: (i) almost all synthetic inputs are prohibited, and (ii) 'soil building' crop rotations are mandated.

The basic rules of organic production are that natural inputs are approved and synthetic inputs are prohibited, but there are exceptions in both cases. Certain natural inputs determined by the various certification programmes to be harmful to human health or the environment are prohibited (e.g. arsenic). As well, certain synthetic inputs determined to be essential and consistent with organic farming philosophy are allowed (e.g. insect pheromones). Lists of specific approved synthetic inputs and prohibited natural inputs are maintained by all certification programmes. Many certification programmes require additional environmental protection measures in addition to these two requirements. While many farmers in the developing world do not use synthetic inputs, this alone is not sufficient to classify their operations as organic.

8.4 Fair trade

According to the major four international NGOs involved in fair-trade,[2] fair trade is a trading partnership, based on dialogue, transparency and respect, which seeks greater equity in international trade. It contributes to sustainable development by offering better trading conditions to, and securing the rights of, marginalized producers and workers – especially in the South. Fair trade organizations are engaged actively in supporting producers, awareness raising and in campaigning for changes in the rules and practice of conventional international trade.

There is a variety of fair trade standards developed by a number of NGOs. In the agricultural sector, the most widespread system is that of the Fairtrade Labelling Organizations International (FLO), an international NGO based in Germany. FLO comprises 20 national fair trade labelling NGOs, mostly from developed countries in Europe, North America, Asia and Oceania. FLO's member organizations work with small producers and farm workers to increase their security and economic self- sufficiency, and empower them in their own organizations. The FLO system relies on certification, i.e. compliance with the FLO standard is verified by a third party that does not have an interest in the business transaction. FLO is responsible

[2]The International Fair Trade Association (IFAT), FLO (Fair trade Labelling Organizations International), NEWS! (Network of European Worldshops) and EFTA (European Free Trade Association).

developing the standard and supporting producers, but the fair trade certification is carried out by a separate organization, FLO Cert, a not for profit NGO. The FLO fair trade system guarantees agricultural producers a minimum price and a price premium on product sales.

FLO has developed and regularly updates detailed standards for crops. To obtain certification, producer associations must function in a democratic manner. There are also rules on how the fair trade premium has to be spent and requirements for the protection of the environment. For plantations, there are a number of requirements related to labour rights: treatment of workers, freedom of association and collective bargaining, workers' housing and sanitation; workers' health and safety; and no child or forced labour. In addition, the producer must comply with the environmental and social laws in the producing country and demonstrate continual improvement in annual inspections (audits).

Other fair trade certification systems have emerged recently. They have been developed by private certification bodies, notably Ecocert (France) and IMO (Switzerland). The International Organization for Standardization has debated the relevance of developing a standard for fair-trade, but no decision has been taken so far. To date, no government (except France) has undertaken to regulate fair-trade. This means that the term 'fair trade' is not legally protected and can be used by anyone under any trading conditions. However, the labels used by fair trade NGOs such as those listed above are private trade marks protected by law.

It should be noted that a number of *alternative trading organizations* (ATOs) import foods under fair trade principles although they do not belong to the FLO system. They usually do not use certification, but instead themselves monitor the compliance of their suppliers with their standard (second party verification). Some of these organizations have existed for several decades, well before the creation of FLO, and import significant quantities of foods. Examples include GEPA (in Germany), Oxfam VW (in Belgium) and the Alter Trade Group (in Japan). This chapter uses the term 'Fairtrade' created by FLO to designate those fair trade products which are certified under the FLO system.

8.4.1 Rainforest Alliance

The Rainforest Alliance is a not-for-profit NGO based in the United States and Costa Rica dedicated to environmental conservation. It is a founding member of the Sustainable Agriculture Network (SAN), a group of non-governmental organizations working for environmental conservation and development. The Rainforest Alliance certification aims to promote good farm management practices for natural resource conservation and to improve worker conditions and community relations and environmental management. In collaboration with the producers, SAN has developed standards for fruits, coffee, tea, cocoa, fern and cut flower production.

The environmental requirements of the standard include: conservation of forests, streams and wildlife; soil and water management; storage, transport and application of agrochemicals; integrated pest management; criteria for waste management; and a farm management plan that integrates the environmental and social standards. Some of the criteria, particularly on the social aspects, require compliance with national legislation and internationally recognized conventions.

The Rainforest Alliance certification for farms is carried out by an international certification company, Sustainable Farm Certification International. After the initial audit, there is an inspection every year. All farms must achieve a minimal level of compliance with SAN standards and demonstrate continual performance improvements to maintain certification. The producer pays the cost of farm inspections and an additional annual fee to SAN that depends on the area of land to be certified. The certification mark is mostly used in promotional activities, but is increasingly being used directly on products as well.

The Rainforest Alliance certification generally requires higher environmental and social standards in relation to conventional production methods. An important characteristic is the use of a point system that allows for certain flexibility. Also the certification allows for the use of agrochemicals under certain guidelines. These characteristics may be important for producers in particular farming situations. The Rainforest Alliance does not guarantee a price premium but claims that most certified producers can negotiate a price premium ranging between 0 and 30 per cent because of increased quality and widespread recognition for its label (Liu, 2009). Information on premiums is difficult to obtain. Whether certification will give a financial benefit to the producer may depend on market recognition, and the negotiations between buyers and sellers.

In February 2009, the Rainforest Alliance's website reported that it had certified 31 158 farms in 19 countries for a total area of 527 090 hectares. The Alliance estimates that 1 250 000 farmers, farmer workers and their family members directly benefit from the programme (see www.rainforest-alliance.org).

8.5 Main markets for labelled foods

There is ample evidence that sales of foods with environmental and social labels have expanded rapidly since the late 1990s. However, there is a lack of official data on the volumes and values of sales, as national agricultural census data and official trade statistics usually do not distinguish between certified and non-certified products. In the case of organic products, a few market research firms and NGOs have started publishing data. In the case of the Fairtrade standard, FLO and its member organizations monitor the marketed volumes and (sometimes) values. Data on total Rainforest Alliance product sales are not available, but this organization provides some estimates for the volumes of particular commodities (e.g. coffee, bananas). In order to guide decision-making and policy formulation, it will be necessary to establish systems for collecting data on the markets for certified products in a more systematic manner.

Developed countries are the main markets for certified products with more than 95 per cent of sales, but there is a rapid increase in some other countries such as Brazil, Argentina and China (Liu *et al.*, 2004). In Europe, Western European countries account for the bulk of the market (more than 90 per cent), but increases have occurred in Central Europe (Czech Republic, Slovenia, Slovakia and Hungary). Switzerland has a very high per capita consumption (*idem*). There is a large variation in consumption per capita across the different EU countries, with Germany, United Kingdom and France leading by volume as the most important markets (*idem*).

There is a wide range of environmental and social labels available in the stores of developed countries. This is a positive development, as it gives consumers information and the possibility to choose the products that address their concerns about sustainability. However, the proliferation of labels may also create confusion among consumers, who do not always know what a label guarantees, and there is the risk of deception.

The remainder of this section describes the markets for organic and Fairtrade certified products, which are those for which more complete sales data are available.

8.5.1 Organic-labelled foods

Based on estimates collected from various studies and industry sources,[3] global retail sales of organic-labelled foods were estimated at some US$40 billion in 2006. Few final figures are available for 2007 yet, but Organic Monitor (2009) estimates that sales reached US$46 billion. They have increased four-fold over a decade, growing from approximately US$11 billion in 1997 (Fig. 8.1). Double-digit growth was common for many years, but it has slowed since the second half of 2008 due to the economic crisis.

It is estimated that 98 per cent of the sales of certified organic products take place in developed countries. North America and Europe account for the bulk of retail sales as illustrated in Fig. 8.2. Other sizeable markets are Japan, Australia and New Zealand. Although developing countries presently account for only a fraction of sales, consumption is rising steadily in some of them, in particular in the emerging economies of East Asia (Singapore, Malaysia, China, Republic of Korea) and Latin America (Argentina, Brazil, Chile). In these countries, organic sales are overwhelmingly concentrated in the large cities and purchasers originate from the upper classes.

In terms of market share, private organic labels have become somehow marginalized by the development of governmental standards. In most developed countries, governments have regulated the production, marketing and labelling of organic foods since the 1990s (EU) or early 2000s (USA, Japan). However, private organic standards continue to exist alongside public standards due to consumer preferences. In these cases, the food product is certified to two standards (the public and private ones). The percentage of products bearing a private organic label is unknown.

---

[3] ITC, Eurofood, SÖL, Organic Monitor and other sources.

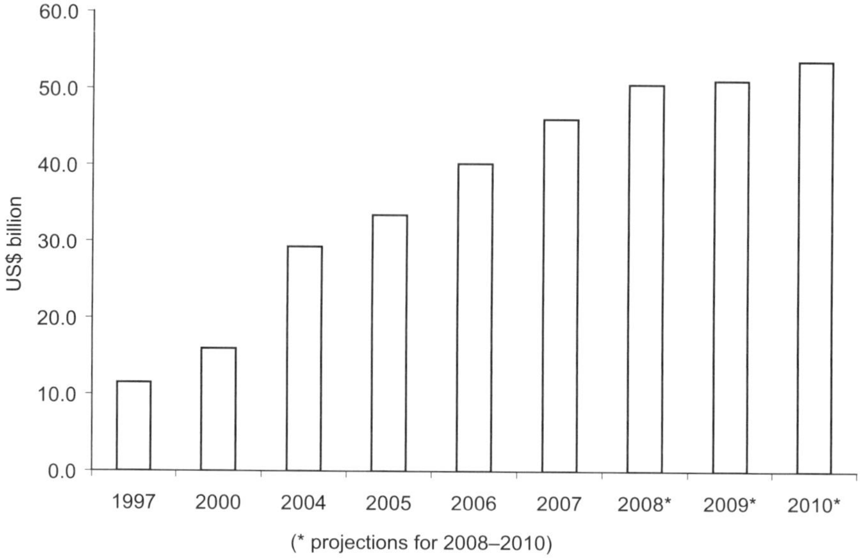

Fig. 8.1 World retail sales of certified organic products (past and projected). Source: Liu (2009).

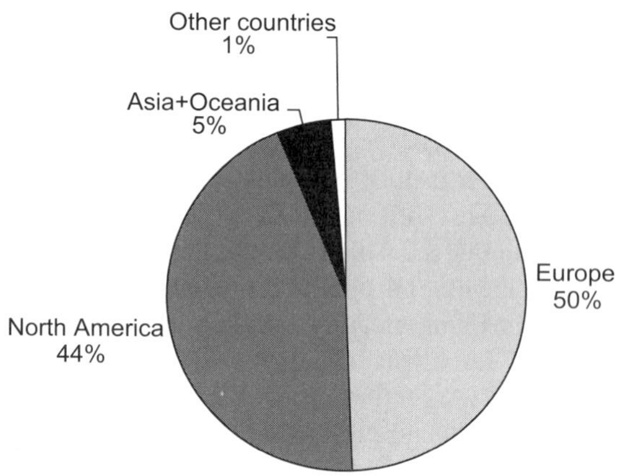

Fig. 8.2 Main markets for organic foods (in percentage of world retail sales in 2006). Source: Liu (2009).

8.5.2 Fairtrade-labelled foods

Global sales of Fairtrade certified foods reached nearly €2.4 billion (US$3.5 billion) in 2007 according to the Fairtrade Labelling Organizations International (FLO, 2008).[4] Sales increased by 47 per cent (in euro terms) over their level of

[4]Since this figure only reflects sales of FLO certified foods and does not include sales by alternative trading organizations, the total market value of fair trade food is slightly higher.

2006, and further growth was recorded in 2008. Tropical products such as tea, cocoa, coffee and bananas enjoyed the fastest growth rates. On average, sales expanded by 40 per cent annually over the period 1997–2007. FLO certified products are available in more than 60 countries. The main markets for fair trade products are the United States, the United Kingdom, France, Switzerland and Germany, accounting for nearly US$2 billion in 2007 (82 per cent of global sales of FLO labelled foods).

By the end of 2007, 632 producer organizations in 58 developing countries in Africa, Asia, the Caribbean and Latin America were certified by FLO (FLO, 2008). FLO estimates that these organizations represent 1.5 million farmers and farm workers and, when counting their families and dependants, overall 7.5 million people benefit directly from fair trade (*idem*). The number of certified producer organizations has trebled since FLO was created in 1997. Some NGOs that do not belong to the FLO system also sell fair trade labelled foods, but the quantities are very small compared to those of FLO labelled foods.

8.6 Benefits and costs for producers

The main incentive that spurs producers to seek certification is the expectation of a price premium. Indeed, some environmental and social labels may have a direct value adding impact by enabling producers to obtain higher sale prices. In developed countries, a substantial share of consumers is willing to pay a price premium for products that can offer guarantees that their environmental, health and social concerns with regard to food production are addressed. Under the pressure of declining commodity prices at the end of the 1990s, many agricultural producers have sought to differentiate their products from those of their competitors by targeting premium market segments. Traditionally, product differentiation has been pursued through improving the physical attributes of the goods, be they visible (e.g. grade, shape, colour, physical integrity, variety, packaging) or not (e.g. taste, acidity, sugar content). More recently, however, farmers and processors have started to differentiate their products on the basis of the production process. Environmental and social standards offer an avenue for such differentiation.

These labels are of particular interest to developing economies where they may help to generate employment, raise export earnings, support small producers, improve food security and resilience to climate change, preserve environmental quality and diversify the local economy. Certification is a strategy for producers and exporters to add value to their products and increase the economic viability of small-scale agriculture. Rising demand for certified products creates new market segments where producers may be able to demand price premiums and secure buyers for their products.

Beside the direct price effect, engaging in the certification process may yield other advantages for food producers. The required traceability and record keeping systems may improve the management of the farm or company. They may help them rationalize production and cut input costs (for example through a more efficient use of agrochemicals). Complying with standards may improve market access through enhanced product quality and improvement in the image of the farm or company. Labour standards may reduce worker turnover, absenteeism and accident and sickness rates, thereby reducing costs and raising productivity. They may lead to better health conditions for farmers and farm workers. Compliance with environmental standards may improve the management of natural resources on which farmer livelihoods depend. They may enhance the farmer's relations with the local community, including its suppliers and lenders. Although they are difficult to quantify in financial terms, these benefits may be significant.[5]

On the other hand, complying with new standards usually entails additional costs for suppliers. Investments are often necessary to upgrade production. Obtaining and maintaining certification is costly, as suppliers have to pay registration and inspection fees. This problem is compounded when farmers produce for different clients requiring different standards. They have to go through several certification processes, which is costly and time consuming. This is one of the negative consequences of the proliferation of certification schemes.

8.7 Case study: organic bananas

8.7.1 Labelling in the banana industry

Environmental and social labels are becoming more widespread in the banana industry (Liu, 2009). Table 8.2 displays estimates of the export quantities of bananas bearing the fair trade, Rainforest Alliance or organic agriculture label. Exports of bananas bearing those labels were estimated at over 2 million metric tonnes in 2007, accounting for close to 15 per cent of global banana exports. The exact value of retail sales is unknown due to the lack of price data, but the global value was likely to approach US$3 billion in 2007.

Table 8.2 Estimated exports and sales of bananas bearing selected sustainable agriculture labels

Standard	Estimated global exports (MT in 2007)	Estimated share of world banana exports (% in 2007)	Estimated sales in 2007 (US$ million)
Organic agriculture	310 000–330 000	2.2	800
Fair-trade	250 000–260 000	1.7	450
Rainforest Alliance	1 500 000–1 700 000	11	1 800
Total(*)	2 000 000–2 200 000	14.5	2 900–3 000

(*) the total is less than the sum of the rows due to multiple certification
Source: Liu (2009).

[5]For a literature review of the impacts of certification in agriculture see Dankers (2003) and Cuffaro and Liu (2007).

The bulk of certified bananas are exported from developing countries (in particular Latin America and the Caribbean) to developed countries. Among the latter, Europe and North America predominate, accounting for some 90 per cent of imports. Japan follows at a distance, with the Philippines and South America as its primary suppliers. Europe imports organic and fair trade bananas from Latin America, the Caribbean and West Africa. North America imports organic bananas from Latin America.

8.7.2 Trade and markets

Developed countries account for the bulk of imports of certified organic bananas. Europe, North America and Japan together represent 99 per cent of imports (Liu, 2009). Europe alone accounted for over half of world imports in 2006 (Fig. 8.3). The retail value of organic banana sales worldwide was estimated at US$800 million in 2007 (Liu, 2009).

World exports of certified fresh organic bananas were estimated to exceed 300 000 metric tonnes in 2007, accounting for over 2 per cent of global sweet banana exports. As can be observed in Fig. 8.4, exports have risen nine-fold since 1998. The rise was particularly strong between 2004 and 2007 for two reasons. First, in 2005 and 2006 production in the Dominican Republic recovered from the damage caused by bad weather in 2004. Second, Ecuador and Peru raised their shipments markedly over these years.

The production of organic bananas shows a strong concentration in the Latin American and Caribbean region. Although no recent figures for production are available, it can be estimated based on the export quantities and certified areas that close to half a million tonnes were produced in 2007. The world's largest exporters of organic bananas are Ecuador, the Dominican Republic, Peru and Colombia.

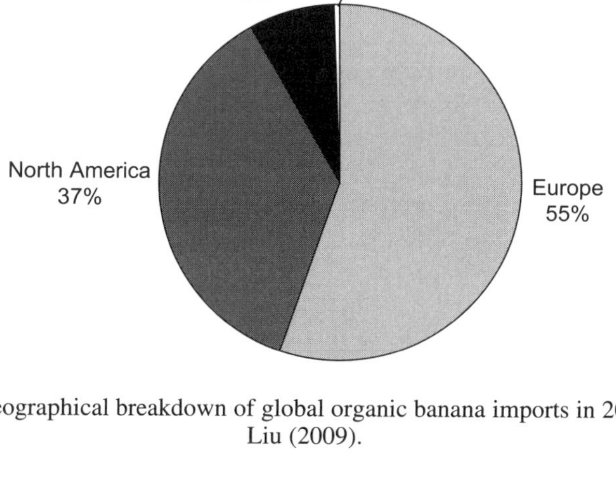

Fig. 8.3 Geographical breakdown of global organic banana imports in 2006. Source: Liu (2009).

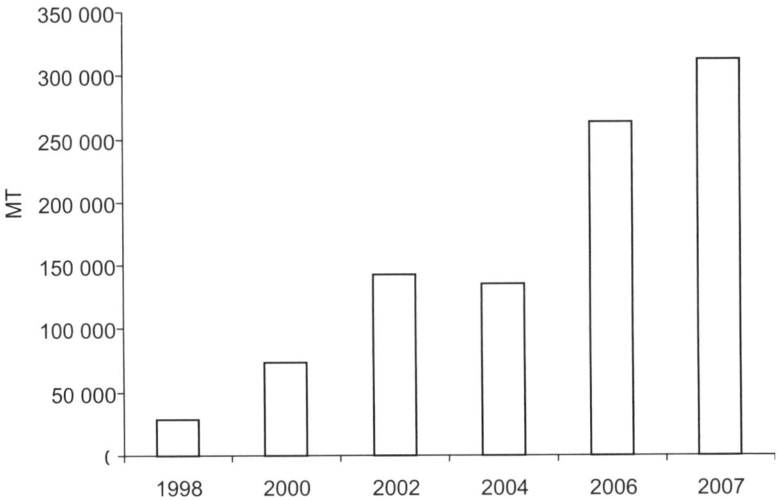

Fig. 8.4 World exports of fresh certified organic bananas 1998–2007 (metric tonnes). Source: Liu (2009).

Ecuador's share has soared in the past three years, and in 2007 it accounted for over 40 per cent of global supply.

8.7.3 Benefits and costs for producing countries

Recent analyses (Liu, 2009; Roquigny *et al.*, 2008) suggest that there is a price premium at exporter level for developing countries shipping certified bananas. The price premium results from consumer preference for organic foods. The size of the premium varies substantially across producing countries, over time and depending on the chosen standard.

However, the higher FOB prices for organic bananas do not necessarily translate into net gains for exporting countries, as they also reflect higher costs. The strict technical requirements of organic agriculture standards may decrease yields and raise production costs, especially during the transition period. The effects on yields and costs depend on how intensive production was before conversion. Traditional low input farmers may expect yield gains from conversion to organic agriculture methods (Dankers, 2003). However, higher yields are usually accompanied by higher production costs, mainly in the form of increased labour demand. In the cases of conversion from high-input production systems, initial yield declines are often observed, usually recovering to levels slightly below the original conventional yields. Effects on production costs per hectare depend on the agro-ecological context, farm structure and size and farmer skills. Organic cultivation of bananas requires technical skills and investment in time. Some tropical diseases, in particular Black Sigatoka, are difficult to combat with organic methods. They require constant monitoring and labour.

Compliance with the strict environmental requirements of organic standards may improve the management of natural resources on which farmer livelihoods depend. They may enhance the farmer's relations with the local community, including its suppliers and lenders. Although they are difficult to quantify in financial terms, these benefits may be significant. More broadly, organic farming generates a wide range of public goods including the preservation of natural resources (water, air, soil, biodiversity), maintaining amenities and reducing health problems caused by agrochemicals.

It has often been observed that the quality requirements of the new organic market are higher than for the former conventional market. In a case study of the Dominican Republic (Damiani, 2002), price premiums were apparently not sufficient to justify the necessary investments to significantly improve the quality of organic bananas grown by small scale producers, and it was difficult for them to compete in the increasingly demanding international organic market.

Certification costs are a key determinant of the profitability of organic banana cultivation. For small growers, the use of group certification involving an internal control system is important to reduce these costs. Developing internal control systems requires institutional changes in farmer organizations. Group certification can be achieved in two distinct ways. First, through associations, with farmers participating actively in decision making and monitoring, in which cases the certificate is owned by the association. In the second system, the exporter organizes and pays for the certification.

Case studies suggest that a relatively small share of the price premium paid by consumers accrues to the exporting country (Roquigny *et al.*, 2008; Liu, 2009). Most of it is captured by downstream operators in the import market. While the premiums found generally exceeded one dollar per kg at retail level, they only ranged between 5 and 20 US cents per kg at exporter level (accounting for between 5 and 18 per cent of the premium at retail level) depending on the exporting and importing countries examined (Liu, 2009). In percentage terms, the premium varied along the supply chain and was at its maximum at the wholesaler/distributor level. Analysing the evolution of prices along the supply chain, it was found that retailers capture the largest share of the retail price. In the above-mentioned cases, this share ranged between 40 and 48 per cent. This situation highlights the strong bargaining power of large-scale retailers.

In sum, organic labelling enables banana exporters to obtain a higher price, but market distortions prevent them from reaping the full benefits. This reduces the returns to investment in organic production and the incentives for growers to adopt this standard. Also, by generating high retail prices the distortions impede the expansion of the markets for certified bananas. In order to limit market distortions and reap the full benefits of organic labelling, grower organizations should strive to establish short marketing chains on which they can have a sufficient degree of oversight and control. Banana growers should organize in sufficiently large enterprises so that they can reach a critical mass of supply and invest in the necessary facilities to perform the functions of collecting, transporting, packaging and exporting. They must increase the efficiency of management, rationalize production and achieve scale economies. Where possible, they should try to obtain a stake in import companies in order to have a greater say on the distribution of profit although, in practice, the lack of capital makes it difficult. A more realistic solution in the short run is to market through the fair trade distribution channels. Empirical evidence suggests that the double labelling organic and fair trade ensures better prices for growers.

8.8 Case study: fairtrade-labelled coffee

8.8.1 Trade and markets

Coffee is by far the most important fairtrade-labelled product and sales of fair trade certified coffee have grown considerably in the last decade (on average +20 per cent per year since 2002). FLO, (2008) indicates that sales of Fairtrade coffee worldwide reached 62 200 metric tonnes in 2007, up 19 per cent from 52 000 metric tonnes in 2006 (Fig. 8.5). North America has become a leading market, accounting for nearly half of this volume. The fair trade coffee market in the United States has grown considerably in recent years, although growth has slowed since 2007.

Fairtrade coffee accounts for some 2 per cent of the total US green coffee imports. TransFair USA estimates that the retail sales of Fairtrade coffee in the United States reached US$837 million in 2007 (Fig. 8.6), up from US$730 million in the previous year (+15 per cent). It calculates that Fairtrade coffee represents nearly 4 per cent of the US retail market value. The number of firms (roasters and

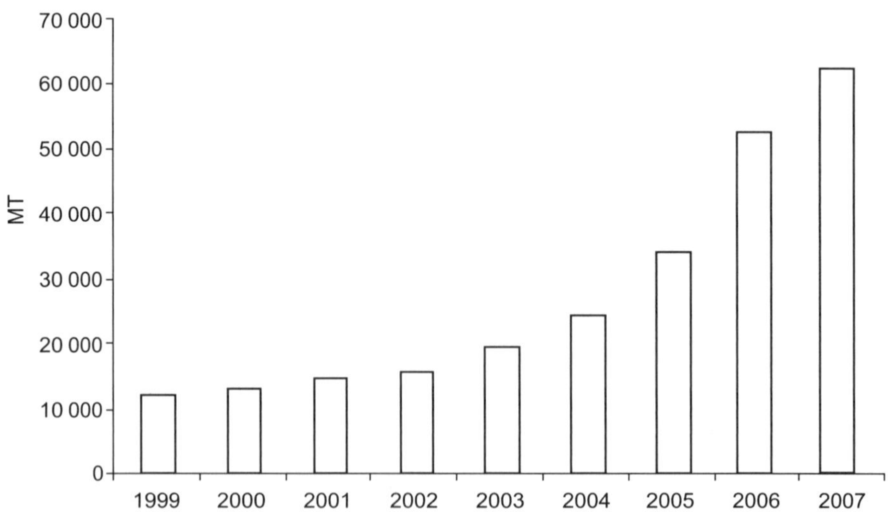

Fig. 8.5 Quantities of FLO-certified coffee sold worldwide 1999–2007 (in metric tonnes). Source: FLO (2008).

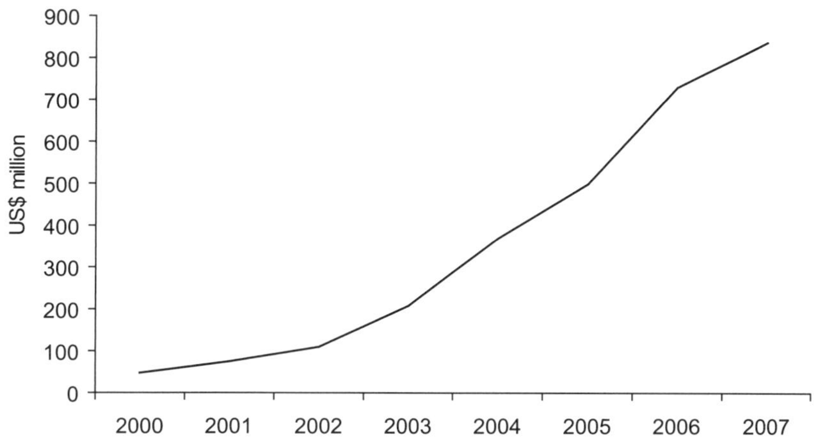

Fig. 8.6 Estimated retail sales value of Fairtrade coffee in the United States. Source: TransFair USA (2008).

importers) licensed by TransFair in the United States has risen steadily since 1999 to 487 firms in 2007.

There is considerable overlap of the organic and fair trade coffee markets. In 2006, approximately 78 per cent of the fair trade coffee sold in the United States was also certified organic while in Canada and the world this reached near 50 per cent on average. This reflects a tendency toward double and even triple certifications; a trend with challenging implications for producers.

Fair trade coffee was exported by 28 countries in 2007 (TransFair USA, 2008). The largest fair trade suppliers were Peru, Mexico, Nicaragua, Indonesia, Ethiopia, Guatemala, Colombia, Uganda and Brazil.

8.8.2 Benefits and costs for producers

Benefits

According to FLO (2008), an estimated 700 000 small coffee farmers directly benefit from fair-trade coffee sales. Most belong to one of the 270 organizations of coffee producers that were certified by FLO in 2007.

The FLO system guarantees a Fairtrade Minimum or floor price that is based on the estimated cost of sustainable production. The minimum price ranges from US$1.01 to US$1.21 per pound depending on the type of coffee and the country of origin (Table 8.3). When market prices rise above the minimum, i.e. US$1.21 for many washed arabicas, a small additional premium is paid.[6] For many years that additional premium was US$0.05 per pound, but in June 2007 it was raised to US$0.10 per pound. The premium is intended for use by cooperatives for social and economic investments at the community and cooperative level.

[6]For arabica coffees (representing the majority of fair trade certification) the market price is determined by the price of the second position 'C' futures contract at the InterContinental Exchange (ICE).

Table 8.3 FLO minimum prices for coffee in 2007 (US cents per pound FOB)

Type of coffee	Central America, Africa, Asia	South America and Caribbean
Washed Arabica	121	119
Non-washed Arabica	115	115
Washed Robusta	105	105
Non-washed Robusta	101	101

Source: FLO, 2008.

When the coffee is also certified organic, an extra premium applies. FLO raised this extra premium by US$0.05 per pound to US$0.20 in 2007. The increase reflects the higher costs of organic production and compliance and also serves as an incentive for greater environmental sustainability.

FLO estimates that the fair-trade system earned farmers an extra income of some €41 million (US$57.4 million) in 2006. This sum represents an average of more than US$200 per farmer above what they would have earned selling on the conventional market. TransFair USA (2008) estimates that the quantities sold in the United States alone generated an additional income of nearly US$19 million for 122 farmer cooperatives in 23 countries.

The first Fairtrade minimum prices for coffee were established by Max Havelaar, a not-for-profit NGO based in the Netherlands, in 1988. Max Havelaar is a founding member of FLO. This system proved very beneficial during the price crisis of the early 2000s. Although the fall in conventional coffee prices caused considerable hardship for small coffee growers across the developing world, the price obtained by Fairtrade growers was often above the international market price (Fig. 8.7). In October 2001, when the market price fell to a record low of US$0.45 per pound, the price of Fairtrade coffee was 180 per cent higher. Recently, as market prices have stayed above the US$1.00 range, the relative premiums for Fairtrade coffee have been more modest (Giovannucci *et al.*, 2008). As such, there are questions about the extent to which producers want to continue with the certification when the price differential is small. For many that do continue there are likely to be two reasons: (i) having a longer-term vision of the cyclical nature of commodity pricing, and (ii) recognizing the other benefits of fair trade.

Indeed, although farmer cooperatives often decide to seek fair trade certification because of the guaranteed price premium, case studies (Dankers, 2003) show that other benefits derived from the fair trade system may be more significant in the long run. The success in self-organization seems to be far more important, resulting in better bargaining positions, better credit worthiness and economies of scale. The fair trade system contributes to these organizational successes through capacity building, an initial guaranteed market, linkages with the international market and learning by doing in exporting. In addition, fair trade contributes to quality improvements. The labour criteria of fair trade standards may reduce worker turnover, absenteeism and accident and sickness rates, thereby reducing costs and raising productivity. They may lead to better health conditions for farmers and farm workers.

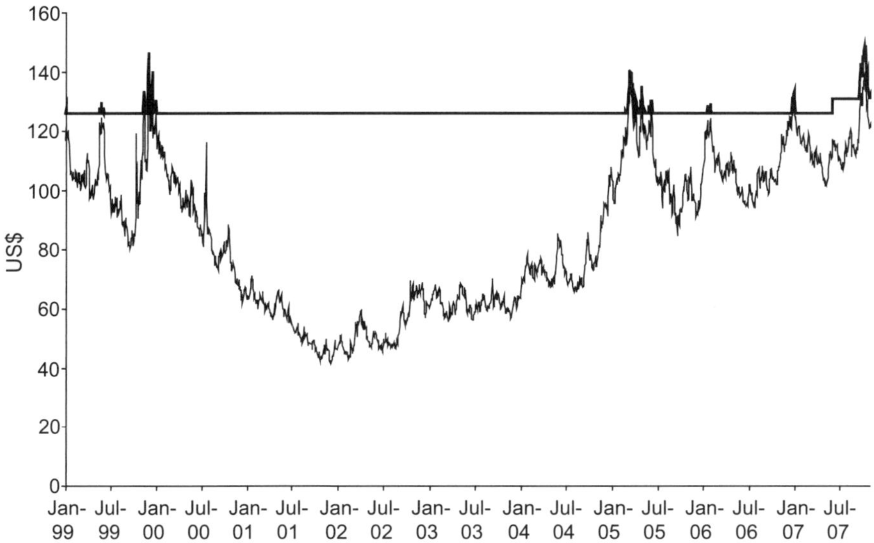

Fig. 8.7 Fairtrade price compared to NYBOT/ICE 'C' coffee price. Source: Giovannucci *et al.* (2008) quoting M. Quinlan Transfair USA based on NYBOT/ICE 'C' market prices.

Costs

The main costs entailed by fair trade derive from the need for farmer groups to modify their internal organization and workings. Similarly as in organic agriculture, Fairtrade certification requires institutional changes in farmer organizations to develop internal control systems. Some organizational changes such as the need for holding general assemblies more frequently, record keeping and hiring independent accountants are likely to raise overhead costs. Yet, there are reasons to believe that growers selling their coffee under the FLO system obtain benefits that more than offset these costs. First, FOB prices tend to be higher and there is a relatively good price transmission from the exporter to the grower, as many Fairtrade groups export directly. When this is not the case, the FLO system aims to ensure that the exporter's margin is not excessive. Second, FLO has a special fund that may partly subsidize the cost of certification at least in the first years.

8.9 Conclusions

The number of environmental and social labels used in the food markets of developed countries has increased markedly over the past 15 years. There is a wide range of labelling schemes, each with its own objectives, scope and approach. Although this development gives producers and exporters more choice, it may also create some confusion among consumers, who do not always know what a label guarantees and to what extent its claim to sustainability can be trusted. From the producing country's perspective, the most interesting labels are those that generate a price premium at producer level and public goods. There is evidence that organic agriculture and fair-trade lead to higher prices for producers and exporters, although a large share of the extra price paid by consumers remains with downstream market operators, in particular retailers.

The market for products labelled as fair-trade and organic has expanded considerably since the mid-1990s and these products are now commonly found in the supermarket chains of developed countries. Growth has slowed since 2008 due to the economic crisis, but sales are expected to continue rising as an increasing number of consumers adopts sustainable modes of consumption. In order to guide decision-making and policy formulation, it will be necessary to establish systems for collecting data on the markets for certified products in a more systematic manner.

Consumers increasingly expect that the foods they purchase address all the dimension of sustainability. Consequently, products bearing multiple certification labels (e.g. organic and fair-trade) have the best market prospects. However, small-scale farmers will need more public support to adapt to the technical challenges and meet the extra costs of some of the standards. In particular, governments and development agencies should support farmer organizations so that they can establish effective systems for quality control and marketing, and provide technical support to their members. Also, more collaboration and coordination among labelling organizations is desirable to limit the burden that multiple certification puts onto suppliers.

8.10 References

CAC (1999) *Guidelines for the Production, Processing, Marketing and Labelling of Organically Produced Foods.* Codex Alimentarius Commission, FAO/WHO, Rome, Italy.

CUFFARO N AND LIU P (2007) *Technical regulations and standards for food exports: trust and the credence goods problem.* Commodity Market Review 2007–2008, FAO, Rome, Italy.

DAMIANI O (2002) *Small farmers and organic banana production in the Dominican Republic.* Office of Evaluations and Studies, IFAD, Rome, Italy.

DANKERS C (2003) *Environmental and social standards, certification and labelling for cash crops.* FAO, Rome, Italy.

FAO (2007) *Private standards in the United States and European Union markets for fruit and vegetables.* FAO Commodity Studies n.3, Rome, Italy.

FLO (2008) *Annual report 2007.* Bonn, Germany.

GIOVANNUCCI D, LIU P AND BYERS A (2008) Adding Value: Certified Coffee Trade in North America. In Liu P (ed.) *Value-adding Standards in the North American Food Market – Trade Opportunities in Certified Products for Developing Countries.* FAO, Rome, Italy.

LIU P (2009) *Certification in the value chain for fresh fruit: the example of banana industry.* FAO, Rome, Italy.

LIU P, ANDERSEN M AND PAZDERKA C (2004) *Voluntary Standards and Certification for Environmentally and Socially Responsible Agricultural Production and Trade.* FAO Commodities and Trade Technical Paper 5, Rome, Italy.

ORGANIC MONITOR (2009) *Organic Monitor gives 2009 Predictions.* www.organicmonitor.com. London, United Kingdom.

ROQUIGNY S, VAGNERON I, LESCOT T AND LOEILLET D (2008) *Making the rich richer? Value distribution in the conventional, organic and fair trade banana chains of the Dominican Republic.* Paper No. 68 presented at the Fair Trade International Symposium, CIRAD, Montpellier, France.

TRANSFAIR USA (2008) *Almanac 2007.* Oakland, CA, United States.

9

Geographic origin and identification labels: associating food quality with location

Emilie Vandecandelaere, Food and Agriculture Organization of the United Nations (FAO), Italy

Abstract: Agricultural and food products differ from others by some characteristics, qualities or reputation resulting essentially from their geographical origin. This specific quality can be promoted with a designation or 'label' referring to the origin location – the geographical indication (GI). There are different motivations for implementing and protecting GIs as recognized intellectual property rights. Indeed, GIs' implementation can add value to origin-linked quality products and so improve livelihoods of rural households. When correctly implemented and managed, they can be a tool for rural development by contributing to local resources preservation and strengthening the organization of local stakeholders. GIs' implementation is a twofold approach: based on voluntary action by producers to define the product's characteristics collectively and to produce the product in accordance with these specifications or code of practice (CoP), GIs can be recognized and registered by public authorities.

Key words: geographic origin and identification labels, associating quality with location.

9.1 Introduction

Some food products are labelled with famous geographical names or indications linked to their place of production (Fig. 9.1). This type of information is therefore not just an indication of source,[1] but refers to a specific quality and reputation due

[1] 'Indication of source' refers to a sign that simply indicates that a product originates in a specific geographical region, in particular some countries, such as 'Made in Germany', 'Product of the USA' or 'Swiss Made'.

Fig. 9.1 Product GI logos: (a) Chivito Criollo del Norte Neuquino Consejo Regulador de la Denominación de Origen (Chivito (baby goat) from the Neuquen region – Argentina). Reproduced with permission of the Counsel of Denomination of Origin (b) Darjeeling Tea, India: Darjeeling logo – Registered intellectual property of the Tea Board of India. Reproduced with permission from the Tea Board of India (c) Le Gruyère – Switzerland: Le Gruyère AOC Switzerland, the true Swiss raw milk tradition. Reproduced with permission of Interprofession du Gruyère (d) Idaho Potatoes: a collective trademark registered by the Idaho State. Reproduced with permission of the Idaho Potato Commission (USA).

to the local natural and human resources of a delimited area. Some of these are internationally well known such as Champagne wine from France or Parmigiano-Reggiano cheese from Italy, while others have only a national or local reputation.

In fact, concerning the wine sector in particular, geographical indications have long been in existence. The first references can be found in the Bible, where wine of Samaria, wine of Carmel, wine of Jezreel or wine of Helbon are mentioned, and references continued throughout Antiquity and the Middle Ages. Then official rules were implemented. Regarding cheese products, Roquefort cheese was first mentioned in historical records in 1070. Then in the 15th Century, King Charles VI of France granted the villagers of Roquefort the exclusive rights to produce Roquefort cheese that should be matured in nearby communal caves, and thus producers of counterfeit Roquefort risked punishment. With regard to wine, the oldest regulation referring to classified vineyards and controlled appellation took place in the 18th Century. Chianti in Italy, Port wine produced in the region of the Douro Valley (Portugal), and the Tokaj-Hegyalja Habsburg Empire (in modern day Hungary).

9.2 Labels on quality linked to geographical origin: rules and diversity in the international context

9.2.1 Quality linked to geographical origin and geographical indications

Some food products can be promoted with a designation or label referring to the origin which is very often used by local actors and consumers to identify some particular and well-known food (FAO, 2004). This designation referring to the origin then differentiates such products from others in the same category based on some specialized characteristics, quality or reputation essentially due to their geographical origin.

This specific quality can be attributed to the history of the product and to a distinctive character linked to natural and human factors such as soil and climate, local know-how, or traditions. In this sense, the 'terroir' demonstrates the interaction between the physical (natural) and human factors built up over time and leads to uniqueness, identity and value of the products.

Geographical indication (GI) is a place or country name that identifies the origin, quality, reputation or other characteristics of products. A GI signals to consumers that the goods have special characteristics due to their geographical place of origin. 'Appellation of origin' represents a more restrictive category of GIs as: geographical designations of products whose quality and characteristics are due exclusively or essentially to the geographical environment, including both natural and human.

GIs are different from an 'Indication of source' reference which simply indicates that a product originates from a geographical region or particular country, such as 'Made in Germany', 'Product of the USA' or 'Swiss Made', without referring to the product quality.

The use of geographical indications calls therefore for a definition of the specific quality and a demonstration of its link to the geographical origin. The definition of the product and the local rules that are followed by the value chain actors in the production of a GI product are described in a document called code of practice (see Section 9.4.1). This code of practice should give both clear guidance to local producers and quality assurance to consumers.

A geographical indication associates a specific product with a territory and therefore its related code of practice and encompasses three main elements:

- a defined geographical area of production;
- specific quality of the product due to specific characteristics of production and processing;
- a name and reputation that differentiate the product from others.

Different types of geographical indication exist: it can be a geographical name that becomes the name of the good such as Champagne or the wines of Bordeaux. Alternatively, the geographical word can be linked to the common name of the good, as for example: Coffee of Colombia or Chivito (baby goat) of Neuquén in Argentina, or Limon of Pica in Chile. The name or symbol – with or without the common name of the good – can refer to a place and its local people without bearing a geographic word such as, for example, Tequila in Mexico, Feta cheese in Greece or Basmati rice in India. Additional associated characteristics can also be considered as geographical identifiers, such as: images of famous places like mountains or monuments, flags, images of specific objects, folkloric symbols, etc., as well as a specific traditional shape or appearance of the product, such as a specific packaging or a common element of the label.

Because of the reputation and value attached to the local name, origin products can be subject to imitations and counterfeiting, thus misleading consumers, by the use of the GI for products that do not conform to the code of practice. These unfair practices may endanger the reputation of the product and the functioning of the value creation process or hinder beneficial outcomes to the local community. It is therefore necessary to protect geographical indications and to ensure conformity with the code of practice in order to avoid unfair production and commercial practices, guarantee the quality of the product and of the geographical origin, and foster consumers' confidence. This regulatory process is also useful to enhance coordination and cohesion among GI producers.

9.2.2 Legal and institutional framework for geographical indications

Historically, some official recognition has existed since the Middle Ages in Europe. Today, various legal instruments are available to protect GIs depending on the country. These include:

- national laws on business practices relating to the repression of unfair competition or the protection of consumers either in general terms or more specifically in regard to such matters as the labelling, certification and agricultural control measures, etc;
- regulation of GI registration under intellectual property rights: specific geographical indication laws and trademark laws, with different categories depending on the countries.

International instruments are quite recent and consider GI as intellectual property rights. They include: the Paris Convention for Protection of Industrial Property, the

Madrid Agreement for the Repression of False or Deceptive Indications of Source on Goods, the Lisbon Agreement on the Protection of Appellations of Origin and their Registration, and TRIPs (Trade-Related Intellectual Property Rights) (see Chapter 3).

For example, Champagne enjoyed an appellation control by virtue of legal protection as part of the Treaty of Madrid (1891) that aims at 'the repression of false or deceptive indications of sources on goods' (WIPO, 1891). The 1958 Lisbon Agreement on the Protection of Appellations of Origin and Their Registration offers the strongest protection for GIs (WIPO, 1958). It defined the Appellation of Origin as the name of products whose 'quality and characteristics are due exclusively or essentially to the geographical environment, including natural and human factors' (WIPO, 1958).

More recently, geographical indications were defined as such in 1994 within the Trade-Related Intellectual Property rights (TRIPs) Agreement of the World Trade Organisation (WTO) as: 'indications which identify a good as originating in the territory of a Member, or a region or locality in that territory, where a given quality, reputation or other characteristic of the good is essentially attributable to its geographical origin'.[2] So, a GI also indicates that a product originates in a specific region, but implies a specific quality due to the geographical origin (WTO, 1994).

The TRIPs agreement requires that the WTO Members provide the legal means to prevent the misleading use of GIs, including when the origin indicated on a product is other than its true place of origin, or when the use of a GI in some way constitutes an act of unfair competition. Countries can meet these obligations through a variety of legal tools, either through existing intellectual property laws (collective or certification trademarks if appropriate), consumer protection or competition laws or by enacting a specific legislation dedicated to the protection of GIs and appellations of origin (AO) (*sui generis system*).

In practice, at the national level, there are two main categories of protection under intellectual property rights:

- Public approach through an official recognition and regulation of the name associated to a specific quality product: this type of scheme aims at protecting the real identification of the origin and its link with quality and reputation. It is based on a strong involvement of public authorities with the definition, implementation and enforcement of the scheme. The code of practice is elaborated by private stakeholders, and then recognized by the public authorities. Any producer who can meet the requirements of the code of practice can benefit from the GI. This is the case for Protected Designation of Origin (PDO) and Protected Geographical Indication (PGI) in the European Union, the Geographical Indication and Appellation of Origin in Morocco, the Appellation of Origin (Denominación de Origen) in the Latin American countries who are part of the

[2]Article 22.1 of the trade-related aspects of intellectual property rights (TRIPS) agreement of World Trade Organization (WTO).

Andean countries Community (Bolivia, Colombia, Ecuador, Peru and Venezuela) as well as Brazil and Mexico, the AO and GI in Chile and Costa Rica.
- Private approach through trademark law: Some trademarks can be used by a group of producers (collective or certification trademarks, depending on the national framework). They aim at certifying quality, characteristics, geographical origin and/or a method of production according to the requirements of a self-established regulation. The protection is therefore based on private actions and the membership of the association may be restricted according to the decisions of its members.

Case study 1: *La marca colectiva del queso 'Cotija región de origen' – Mexico*

The Cotija cheese from the Jalmich mountain range in Mexico takes its name from the nearby city of Cotija and is very well known for its high quality throughout the whole country. However, the genuine 'queso Cotija' is threatened by usurpation of the name by cheeses called 'type Cotija' which are produced outside the original production area and have caused the name to become generic. These cheeses are usually industrial (intensive production, no maturation, with filling, etc.) and the taste is very distinctive from the authentic types, but they tend to be cheaper. In order to fully protect the name and reputation of their product, the producers of the typical Cotija cheese applied to the authorities in charge of intellectual property rights to register the product, based on elaborating the code of practice involved in its manufacture. However, because the name 'Cotija' had come to be so widely used, they were unable at that point to obtain the denomination of origin (DO) status which they considered to be the most effective legal protection for Cotija and its reputation. However, they were able to retain the collective trademark 'Cotija Region of origin'.

(adapted from Poméon, 2008).

9.2.3 Importance of labelling and the guarantee system for conformity assessment

Geographical indications help consumers to recognize, through the label, the specific quality linked to geographical origin, but this reference has to be guaranteed. In some cases, particularly in local markets, consumer confidence may be based on the short distance between consumers and producers. But, as the distance between the places of production and consumption widens, a certified and monitored information system must be established both to inform the consumers and to guarantee the conformity of the product with the requirements on the code of practice.

Regarding labelling, in the case of the public scheme, a national or official and common logo often allows consumers to recognize the GIs more easily and to

Fig. 9.2 National GI logo: The Common Swiss logos. The *Association suisse pour la promotion des AOC et des IGP* was set up in Bern in 1999 to associate all the supply chains willing to protect their products with a PDO or a PGI. The aim of the Association is to promote the AOC (PDO) and the IGP (PGI) label in Switzerland to consumers and retailers. The Association encourages the use of its common AOC or the IGP logo by its members so that all the Swiss registered products have the same visual identity to inform the consumers they are AOC or IGP products. http://www.aoc-igp.ch/. Appellation d'Origine Contrôlée, Indication Géographique Protégée.

know that the GI is guaranteed (Fig. 9.2). Those logos became so meaningful in the consumers' minds that it took on the significance of a quality sign thus contributing to creating a ranking system for consumers and so helping them to choose knowingly.

In some cases, national authorities can monitor the integrity of the verification applications for geographical indication. It was the case in France with the French National Institute for Appellation of Origin (INAO); now controls are done by third party organizations under agreement with public authorities.

Verification systems serve to ensure the product is conforming to the CoP (Code of Practice), on a voluntary basis. They may differ among countries or regions, depending on the objectives, type of markets, and the economic, social and cultural contexts (Liu and Vandecandeleare, 2008). In any case, internal control should be ensured by the producers themselves. The different verification systems that can be implemented and combined are:

- *Internal control system (first party verification)*: in which a stakeholder (being part of the GI system) gives a self-declaration of conformity to the code of practice. This can be managed by a local association of stakeholders (producers, local authorities, buyers, etc.) that do their own GI supply chain control. It is, for example, the case for Chivito of Neuquén, which is sold essentially on local and regional markets. The local organization verifies the meat conformity, carried out in practice through the local slaughterhouse.
- *Second party verification system*: involves a trading agent who verifies that suppliers comply with the CoP criteria.
- *Third-party certification system*: involves an independent and external body, without direct interest in the economic relationship between the supplier and

buyer, which provides assurance that the relevant requirements have been followed. Specific certification bodies can be organized with public authorities (fully public certification or joint public and private initiatives). For example, the National Federation of Coffee Growers of Colombia is an independent not-for-profit organization for the collective of over 560 000 coffee growers of Colombia that demanded the recognition of the Appellation of Origin Café de Colombia in Colombia and in the European Union. It has since been accepted and protected in the European Union as a PGI according to EU Regulation 510/2006, the control being made by ALMACAFE, is satisfying the international norms for certification (ISO 65).
- *Participatory guarantee system*: a locally focused quality assurance system based on active participation of stakeholders, internal and external to the GI value chain (even consumers), and built on a foundation of trust, social networks and knowledge exchange. Such an alternative system is entirely realistic in the context of small farms and local, direct markets. It is, for example, the case for the special Gari missè (staple food made from toasted cassava semolina), produced in the village of Savalou (Benin) where the quality control is carried out by the group of women processors. They ensure that the processing rules and marketing practices are carried out; a lack of respect for the rules entails the risk of being expelled from the group.

9.3 The reasons for the development of geographical indications

9.3.1 The consumer demand and social expectations

Consumers are becoming increasingly concerned about how the food they consume is produced. These concerns often relate to the sustainability of the food and how it is produced, as well as its environmental and ethical attributes. Therefore, the origin (country, district, and producer) of foods is very important, especially for consumers who are looking for roots, familiarity and continuity in places, identity and tradition (Wilson and Fearne, 2000). Some consumers may want to support the local or national economy; or they are proud of their cultural identity; or they are sensitive to the specific organoleptic characteristics of these products. These consumers are generally willing to pay more to find such characteristics in the product (Giovannucci *et al.*, 2009).

Moreover, consumers are demanding more guarantee and clarity on labelling, for example in Europe, as shown by various studies (Consumers International, 2004). Problems arise concerning ‘implied green claims’ that give the impression of more traditional production practices without specific substantiation to verify the claim. The survey shows that some consumers were becoming confused by the proliferation of unfamiliar logos and labels which had no direct meaning for them, or that were too difficult for them to interpret with confidence (official guarantee or not).

As the market becomes more global, it appears that there is more importance given to the differentiation of products linked to their origin, not only for export products, but also for locally marketed products in relation to their competition with imported products. This is increasingly the case in developing countries.

In general, the demand for these products increases with economic improvements in societies, urbanization and the degree of integration in the global market. Indeed, regional traditional agricultural and food products are often seen as a response to environmental concerns generated by globalization (transport of products over long distances) and to retailers' driving the supply of food. In the case of transition economies, it can be a response to the rapid modernization process, and the increase of imported processed foods marketed by multinational firms (FAO, 2008).

In developing countries, local products are often very prevalent. With increasing urbanization, origin can be a proxy of quality conveying trust to consumers. These urban populations are keen to eat traditional foods from their place of origin or items that have acquired a certain reputation. There is an increasing demand for such products by immigrants who miss them, leading to some specific channel markets, also known as a 'nostalgic market'.

These consumers' perceptions and expectations explained the development of specific labels related to geographical origin and of specific production practices linked to local know-how, and the importance of credible and officially guaranteed labels for these products. In this sense, consumers are expecting guarantees with regard to:

- origin, method of elaboration, and specificity of the products;
- identification presented on clear and informative labelling;
- traceability.

9.3.2 The producer's view: protecting the local name reputation

Development of such labelling is also driven by producers' motivations, particularly for small and medium size enterprises, which consider GI as a marketing tool in relation to differentiation strategies on market segmentation (niche high quality or popular commodities).

GIs are not only a defensive strategy to protect from usurpation but also correspond to a pro-active strategy to reinforce differentiation of a product, build niche markets, increase added value, or to be a driving force to structure a value chain and meet basic safety, quality and traceability requirements of regulated markets.

Origin-linked quality strategy is also extremely relevant for producers, generally small scale and low capacity, in fragile or marginal agricultural zones to turn constraints into assets so as to access niche markets and increase income levels. Indeed, here the particular production constraints (isolated location with distance and weak transport structures, low level of modernization) can be considered as comparative advantages because they become factors that maintain the traditional and unique characteristics of the product.

Another major key aspect of a GI is the fact that the specification of the product, the code of practice, is locally defined by stakeholders, especially producers, allowing for the placement of producers at the centre of the value chain strategy. This has the added benefit of restoring a decision-making role to local communities, guaranteeing their right to manage their own resources and engage their active participation in value-added food chains.

9.3.3 Rural development: supportive projects and policies

The last and most important driving forces for the development of GI labelling are the supportive project strategies or public policies that can promote the potential of GI as a tool for a sustainable rural development.

These origin quality products can serve as a noteworthy focus for action and local organization. In the framework of agricultural and rural development policies for rural territories, GI products can play an important role in promoting collective action for local management of human and physical resources – becoming a motivation for the organization of actors at the local level. These products can be viewed as a tool for preserving traditions and preventing emigration or firm relocation.

Their contribution to sustainable development can be highlighted according to the three pillars of sustainable development: economic impact, social impact and environmental impact.

- *Economic impact: accessing markets, adding value and benefiting from collective organization.* The setting up of a GI label provides access to new niche markets and/or maintains access to existing ones. The differentiation of the product often leads to premium price and added value and therefore improves the income of local producers. The fact that the code of practice should be set up by the local producers represents an opportunity for a fair redistribution of the added value among the value chain actors. Moreover, the organization of stakeholders around origin quality products strengthens the value chain through a collective approach and a common goal: the territory reputation. Rural activities can therefore be maintained, preventing rural exodus and creating possible diversification activities, especially tourism and gastronomy.
- *Social impact: maintaining activities in remote areas, improving the self-esteem of the producers and preserving the know-how and traditional food.* Collective organization around a GI product strengthens relations between the stakeholders involved in the production process, but also creates a wider social network in the area with other stakeholders, public actors, schools, tourism's actors, etc. The societal recognition of the specific value of the product in relation to local know-how and traditions increases producers' and local inhabitants' self-esteem. This is important, especially for small producers in remote areas where traditional farming system is a way of life and for women who are often involved in the production or processing of these products. Promoting the marketing of origin products can prevent their disappearance and

contribute to food diversity. The link between product, people and place often goes beyond the mere economic aspect making the GI product a cultural or symbolic marker and an element of identity for the local population.

- *Environmental impact: sustainable use of natural resources and biodiversity.* GI production is often based on traditional farming systems that have a lower environmental impact on natural resources than modern techniques and inputs. Consequently, the GI process contributes to preserving natural resources (landscapes, soils, biodiversity) and provides a framework, thanks to the code of practice, for a long-term sustainable use of natural resources. Furthermore, origin products often use traditional and specifically local-adapted species, varieties, breeds and ferments that represent genetic resources. Maintaining these products and production systems could also contribute to maintaining the biodiversity (Larson, 2007).

Therefore GI process and labelling can be a tool for sustainable rural development; it explains the increasing number of GIs in developing countries. Nevertheless, the effects are neither automatic nor systematically positive, depending on how the local process is developed and with whom (participative approach) and on the definition of the product characteristics (what local resources are taken into consideration and how). Indeed, if substantial benefits can be developed, there are also some implementation costs and constraints (Anders and Caswell, 2009): in each case, an assessment should carried out to see whether the favourable conditions are met, at the two levels involved: the local with the value chain and market requirements and the national with the institutional and legal framework.

9.4 Setting up a GI label, a two-level approach

Unlike other specific quality standards, each GI has its own specific code of practice corresponding to the definition of the characteristics of the product linked to geographical origin. The setting up of a GI assumes a twofold approach involving:

- *Local level*: the value chain stakeholders (farmers, producers, processors) and other local actors, public and private, supporting the local process.
- *National institutional level*: the regulatory framework to recognize, support and protect the GIs.

9.4.1 The local level

There are two main phases to be considered by local stakeholders when implementing a quality scheme linked to geographical origin (FAO-Sinergi, 2009).

Setting up the local rules for using the GI, i.e. the qualification of the product

Setting up the rules of the GI requires a precise definition of the product's specific characteristics and the demonstration of the link with the geographical origin that differentiates it from other products of the same category. Even if the process can

Case study 2: ***Turrialba market research and consumer surveys – Costa Rica***

In 2006, a researcher from the University of Santiago di Compostela (Lugo) studied the origins and special characteristics of a cheese produced in Costa Rica. Moreover, he carried out market research and consumer surveys for the registration of 'Queso Turrialba' as a DO (Denominación de Origen). The study allowed for the collection of data and information to support the request and involved surveys of 25 farms and five industrial cheese making units as well as chemical, micro-biological and sensorial analysis. To learn about consumers' opinions and whether they appreciated different aspects of the product, the market research included tasting sessions and testing of images. The market analysis also allowed for identifying the place of purchase preferred by the consumers, their awareness and proof of the product's long-standing reputation. For example, one result that came out from the consumer survey, was that 81.6% of polled consumers agreed with the fact that, among different types of white cheeses, 'Queso Turrialba' was a very distinct and recognizable one.

(Blanco, 2008)

be initiated and supported by external actors, for example NGOs or development public actors, this step requires the active involvement of the legitimate local value chain stakeholders who have to define these aspects, since they are the most knowledgeable about their product and the natural resources involved and the related know-how inherited over generations.

These rules are defined in the document named 'code of practice' (CoP) (or 'product specifications', 'book of requirements' or 'disciplinary document' depending on the context). The code of practice includes the definition of the product (name, characteristics, production and process methods), the delimited area concerned and the guarantee system (control plan with the criteria to be assessed and how). As a consequence, the CoP is a tool for internal coordination (collective rules for a fair competition between producers) and external trust (information on quality guarantee for retailers and consumers).

The definition of product and delimitation of the production area require studies and analyses for which supportive actors are helpful for research, development expertise and networking. Consumer studies can be considered to define the marketing strategy: for which consumers, on which market and for which product presentation.

When the code of practice is elaborated, it can be therefore presented and possibly assessed for the GI registration by public authorities in the case of a public approach.

Management of the quality label

Once the GI label is officially recognized, it still needs to be managed locally. More specifically, this management includes the collective marketing of the produce, the

Case study 3: *Chivito Criollo (baby goat) from the North Neuquino region in Argentina, Patagonia*

Chivito (baby goat) meat comes from a specific local breed that has a particular taste due to specific pastures in the mountainous regions, to its breeding based on transhumance, and on a specific related know-how. The identification of the potential of the product started with the programme for the conservation and improvement of the Neuquén Criollo goat established in 2001 under the auspices of the Instituto Nacional de Tecnología Agropecuaria (INTA) that developed a system for providing improved strains of local ecotypes based on selection criteria proposed by the breeders themselves.

The INTA determined the criteria for quality meat on the basis of what they implemented and the classification of the products. Various workshops were organized with producers and retailers in order to analyse the best tools, not only for protection and promotion on the market, but also with regard to the culture and know-how. A writing committee elaborated the specific rules of production of the Chivito Criollo del Norte Neuquino (code of practice). A total of 150 producers participated in developing the request for 990 of them in total.

In 2006, an association named 'Consejo Regulador Denominación de Origen,' was created for the Appellation of Origin; composed of producers, some intermediaries, and local public authorities in charge of research and development (INTA, municipality, the regional offices in charge of production and social affairs). The dealers were few, but all of them were strongly convinced that they needed to differentiate the product on the market and that they needed to work jointly with breeders.

(Pérez Centeno, 2008)

conformity assessment and the possible evolution of the rules (changes in the code of practice as necessary over time). Collective action should also help to look for continual improvements in sustainability within an extended territorial strategy, by linking with other local economic activities, for example, tourism. Therefore, a GI organization involving all the stakeholders of the value chain is highly recommended, in fact, for all stages of the process, from the setting up and the request for GI's registration to the definition of collective marketing strategy. This collective approach allows lowering the cost of marketing plans and conformity assessments (control) but does not replace individual decision and strategy at the firm level.

9.4.2 The national institutional level

At the national level, public actors play an important role in providing an adequate institutional and legal framework for the recognition and protection of GIs, but also in supporting their implementation in such a way that they contribute to rural development and food diversity preservation. The recognition of the specific quality linked to geographical origin as intellectual property rights is now international, even if there are still a great variety of legal tools in different countries.

Case study 4: *Limón de Pica – Chile*

In the driest desert of the world, Oasis of Pica in Atacama, grows a kind of lemon that is special for its unique scent and high juice content. Such attributes have made this a sought-after product on the market, especially for making spirits. Due to this reputation and the risk of usurpation, a group of producers, supported by several institutions, have proposed to achieve a Origen Denomination for the Limón de Pica (lemon from Pica) in order to protect the good will, prestige of the product, to have better prices and to explore new markets.

In 1999, the cooperative of producers was nominated for a national Contest of the Foundation for Agrarian Innovation of the Ministry of Agriculture, for an initiative with the objective of establishing a differentiation strategy and system for Limón de Pica. Three projects followed, from 1999 to 2007, to provide investments, studies, capacity building and organizational support. The project received support from the Government for building the packing house. Other ad hoc types of support were provided, allowing for an increase of capacity building (from the National Institute of Agricultural Development) and organization of producers to visit and see examples of marketing channels for fruit export (PROCHILE).

With regard to the legal framework, the recognition and protection of geographical indications, appellations of origin for food and agricultural products two laws have been enacted:

- the law18.455 for wine and spirits (alcohols/vinegars),
- the law 19.039 on Intellectual property and the related Decree No. 236 of 25.08.05 of Ministry of Economy Promotion and Reconstruction, for forestry and agrifood products other than wines and spirits.

The law allows any person to request a GI or AO registration as soon as they represent a group of producers, processors or handicraft producers. This request can be done by a national, departmental or local authority on the GI territory. The Ministry of Agriculture, in charge of assessing the request with the code of practice for all forestry and agrifood products, prepares a report and recommendations for the Ministry of Economy, which is in charge of the registration of the GI/AO. It may reject the registration it does not conform.

(adapted from Vandecandelaere, 2008)

Institutional actors are responsible for the evaluation of the producers' requests for recognition, registration and protection of the GIs. On top of the role of assessing and registering GIs under the intellectual property rights at the institutional level, public policies on agricultural and rural development also play an important role in supporting the local process and optimizing the GI system as a tool for sustainable development. Public policies at various levels (local, national and international) can create good conditions and clear rules of the game, for the exploitation of all the potential benefits of the GI product with regard to rural development, by implementing a comprehensive and proactive quality labels policy.

Within this approach, provision of information to the public about the meaning of such labels is important in order to raise consumer awareness and so create favorable market conditions.

9.5 Conclusions

Geographical indication labelling is a way to inform consumers about the specific qualities of a product differentiated by its geographical origin, but it is also a way to address an increasing desire for more information on the production place and to meet social expectations for more sustainable means of production.

Setting up a GI for a food and agricultural product can be a tool for sustainable rural development. For fragile or remote areas, highlighting the specific characteristics due to the origin can be a means to turn production constraints into advantages, because they are the source of the uniqueness and quality of the product.

GI labels benefit both producers and consumers but also, in a much larger sense, the whole community of the territory where the GI is located. Indeed, it is a tool for empowering farmers and producers thanks to collective organization, for improving their livelihoods by allowing them to maintain or access niche markets with added value, for protecting natural resources and promoting local know-how and traditions, and for offering more choices to the consumers, who will also be better informed on the guarantees of the GI products.

9.6 Acknowledgements

The author would like to thank Amélie Le Coent for her help in finalizing the editorial improvements for this chapter.

9.7 References

ANDERS, S. AND CASWELL, A. (2009). 'The benefits and costs of proliferation of Geographical labelling for developing countries', *The Estey Center Journal of International Law and Trade Policy*, Volume 10 Number 1, pp.77–93.

BLANCO, M., IN COLLABORATION WITH GRANADOS, L. (2008). 'Queso Turrialba, Costa Rica'. PRODAR IICA, Organización de la Naciones Unidas para la Agricultura (FAO) y Istituto Interamericano de Cooperación para la Agricultura (IICA). In Riveros, H., Vandecandelaere, E., Tartanac, F., Ruiz, C. and Pancorbo, G. (eds), *Calidad de los alimentos vinculada al origen y las tradiciones en América Latina: estudios de casos*. Lima: FAO–IICA. Available at: http://www.fao.org/ag/agn/agns/Projects_SQP_Santiago/Documentos/Estudios%20de%20caso/Turrialba/Queso_CostaRica.pdf [accessed July 2009].

CONSUMERS INTERNATIONAL (2004). *Green Food claims, An international survey of self-declared green claims on selected food products*, London.

FAO (2004). 24th Regional Conference for Europe. Item Six – Food safety and quality in Europe: Aspects concerning in particular quality, nutritional balance, the importance of agricultural land and cultural heritage ('Terroirs'). Montpellier, France, 5–7 May 2004 (ERC/04/4).

FAO (2008). 26th Regional Conference for Europe. Item Eleven – Promotion of traditional regional agricultural and food products: a further step towards sustainable rural development. Innsbruck, Austria, 26–27 June 2008 (ERC/08/6).

FAO-SINERGI (2009). 'Linking people, places and products. A guide for promoting quality linked to Geographical Origin and sustainable Geographical Indications'. Rome: Food and Agriculture Organization of the United Nations (FAO) and Strengthening International Research on Geographical Indications (SINER-GI). www.foodquality-origin.org.

GIOVANNUCCI, D., BARHAM, E. AND PIROG, R. (2009). 'Defining and Marketing "Local" Foods: Geographical Indications for US Products', *Journal of World Intellectual Property* special issue on GIs.

LARSON, J. (2007). Relevance of geographical indications and designations of origin for the sustainable use of genetic resources. Study commissioned by Global Facilitation Unit for Underutilized Species, Rome.

LIU, P. AND VANDECANDELAERE, E. (2008). Diversité des désignations et labels dans le contexte international. FAO. In *Désignation de denrées alimentaires et bioénergies durables,* Proceedings of the seminar organized by the Swiss Federal Office for Agriculture (FOAG), 11 March 2008, Berne, Switzerland.

Pérez Centeno, M. (2008). 'Chivito criollo del Norte Neuquino' Chos Malal, Neuquén – Patagonia, Argentina. Istituto Nacional de Tecnología Agropecuaria (INTA). Organización de la Naciones Unidas para la Agricultura (FAO) y Istituto Interamericano de Cooperación para la Agricultura (IICA). In Riveros, H., Vandecandelaere, E., Tartanac, F., Ruiz, C. and Pancorbo, G. (eds), *Calidad de los alimentos vinculada al origen y las tradiciones en América Latina: estudios de casos.* Lima: FAO–IICA. Available at: http://www.fao.org/ag/agn/agns/Projects_SQP_Santiago/Documentos/Estudios%20de%20caso/Chivito/Chivito_Argentina.pdf [accessed July 2009].

POMÉON, T. (2008). 'El Queso Cotija, Mexico'. CIESTAAM, Universidad Autónoma Chapingo – Mexico. Organización de la Naciones Unidas para la Agricultura (FAO) y Istituto Interamericano de Cooperación para la Agricultura (IICA). In Riveros, H., Vandecandelaere, E., Tartanac, F., Ruiz, C. and Pancorbo, G. (eds), *Calidad de los alimentos vinculada al origen y las tradiciones en América Latina: estudios de casos.* Lima: FAO-IICA. Available at: http://www.fao.org/ag/agn/agns/Projects_SQP_Santiago/Documentos/Estudios%20de%20caso/Cotija/Queso_COTIJA_Mexico.pdf [accessed July 2009].

VANDECANDELAERE, E. IN COLLABORATION WITH MERY, J.E. (2008). 'Limon de Pica, Chile'. Organización de la Naciones Unidas para la Agricultura (FAO) y Istituto Interamericano de Cooperación para la Agricultura (IICA). In Riveros, H., Vandecandelaere, E., Tartanac, F., Ruiz, C. and Pancorbo, G. (eds), *Calidad de los alimentos vinculada al origen y las tradiciones en América Latina: estudios de casos.* Lima: FAO-IICA. Available at: http://www.fao.org/ag/agn/agns/Projects_SQP_Santiago/Documentos/Estudios%20de%20caso/Limon_Chile.pdf [accessed July 2009].

WILSON, N. AND FEARNE, A. (2000). 'A link between modernity and tradition: the case of several regional food products'. In *The socio-economics of Origin Labelled Products in Agri-food Supply Chains: Spatial, Institutional and Coordination Aspects*, Économie et sociologie rurales, Actes et communications No. 17–2, pp. 277–294, INRA – France. Proceedings of the 67th European Association of Agricultural Economics Seminar, Le Mans, France, 28–30 October 1999.

WIPO (1958). Lisbon Agreement for the Protection of Appellations of Origin and their International Registration. Geneva: World Intellectual Property Organization (WIPO). Available at: http://www.wipo.int/lisbon/en/legal_texts/lisbon_agreement.htm [accessed July 2009].

WIPO (1891). Madrid Agreement Concerning the International Registration of Marks. World Intellectual Property Organization (WIPO). Available at: http://www.wipo.int/madrid/en/legal_texts/trtdocs_wo015.html [accessed July 2009].

WTO (1994). Agreement on Trade-Related Aspects of Intellectual Property Rights (TRIPS Agreement). Geneva: World Trade Organization. Available at: http://www.wipo.int/clea/en/text_html.jsp?lang=en&id=4064 [accessed July 2009].

10

New technologies and food labelling: the controversy over labelling of foods derived from genetically modified crops

Janice Albert, Food and Agriculture Organization of the United Nations (FAO), Italy

Abstract: Consumers' views of genetically modified foods (GM foods) can influence food producers' decisions as to whether to market GM foods or whether to use conventional varieties. Through labelling, supported by certification, consumers could differentiate a GM food from a conventional food. A working group of the Codex Committee on Food Labelling identified seven approaches to labelling of GM foods. GM labelling is mandatory when there are differences in the final product that could have a material effect on the consumer. Several countries require labelling when the final product is different than the conventional product, regardless of whether the difference has no consequences for health. There is little consensus on labelling products which do not contain any GM material but were derived from a GM crop or labelling because of the process of production.

Key words: Codex Committee on Food Labelling, genetically modified foods, mandatory labelling, consumer acceptance of GM foods.

10.1 Agricultural biotechnology and consumers

Since genetically modified (GM) seeds were introduced commercially in 1996, their use has spread to 25 countries in North America, South America, Europe, Asia and Africa (James, 2008). The most common GM food crops are soybean and maize, which are grown primarily in the United States of America (USA), Argentina and Brazil (James, 2008). Conventional soybean and maize, as well as their

GM counterparts, are commonly processed into a range of food ingredients, which are widely used by food manufacturers to produce numerous packaged food products.

With strong competition in the global food market, consumers' views of genetically modified foods (GM foods) can strongly influence the decisions by farmers, commodity dealers, food manufacturers and food retailers regarding whether to produce and market GM foods or whether to use conventional varieties. Yet, a shopper cannot distinguish between foods that are conventional or GM without explicit information since the GM status of a product cannot be determined by sensory perceptions or experience. With the current generation of GM foods, the quality of being derived from GM crops is not revealed even after the product has been consumed. This is known as a 'credence' quality (Jahn *et al.*, 2005). Thus, it is only through labelling, supported by certification, that consumers would be able to differentiate a GM food from a conventional food. Through their purchases of labelled foods in the market, they could indicate whether the quality of being GM is important to them. This could have an impact on the use of GM technology in food production.

10.2 Policy options

With the potentially powerful impact that food labels could have on the future of a new technology, the decisions regarding the labelling of GM food products have been the subject of extensive debate within countries and internationally. As a result of various types of consultations with the biotechnology industry, food producers, scientific societies, consumer associations and environmental organizations, as well as the general public, and consideration of national legislation and existing Codex standards, a number of policy approaches have emerged in different countries. Since these food products are traded worldwide, harmonization of these various labelling options has been a concern of many countries. Internationally, the Codex Committee on Food Labelling (CCFL) began discussions on labelling of GM foods in 1991 and the deliberations are continuing.

10.2.1 Seven approaches to labelling of GM foods

At the 34th Session of CCFL, held in May 2006, a working group was charged with several tasks aimed at resolving the impasse in the deliberations. One task was to consider 'the rationale for Members' approach to the labelling of food and food ingredients obtained through certain techniques of genetic modification/genetic engineering' and to 'identify the current standards, regulations, acts/decrees, etc. among current Members with respect to the mandatory and voluntary labelling of foods and food ingredients obtained through certain techniques of genetic modification/genetic engineering' (CAC, 2007). Another was to identify Members' practical experiences in applying/implementing mandatory and voluntary labelling of food and food ingredients obtained through certain techniques of genetic

Table 10.1 Main approaches to labelling of GM foods

1. Mandatory GM labelling as such of all foods derived from or containing ingredients derived from organisms produced using gene technology (food consisting of, containing or produced from GMOs).
2. Mandatory GM labelling as such of GM foods and food ingredients where novel DNA and/or protein are present in the final food.
3. Mandatory GM labelling as such of GM food where it is significantly different from its conventional counterpart and where GM labelling is required in addition to the significant change.
4. Mandatory labelling of GM foods where it is significantly different from its conventional counterpart and where only the significant difference is labelled, but not the method of production.
5. Voluntary labelling (voluntary labelling guidelines for foods that are or are not products of genetic engineering).
6. No special labelling requirement for bioengineered foods as a class of foods.
7. Labelling requirements under development.

Source: CAC, 2007, p. 2.

modification/genetic engineering (*ibid*). The CCFL working group, comprising of 25 members and the European Union, met in Oslo, Norway in 2007 and in Accra, Ghana in 2008. They reported back to the full CCFL, which reports to the Codex Alimentarius Commission. Based on the comments of governments, the working group identified seven main approaches to labelling of GM foods (Table 10.1).

The approaches in the table are not exclusive; a country's labelling requirements might include several of the listed categories and some products may be exempt or excluded from these regulations. To illustrate the complexity, Table 10.2 provides more detailed descriptions of the policies as explained by Codex delegates.

Within the seven approaches, there are agreements on some key points. All of the approaches require positive labelling when there are differences in the characteristics of the final product that could have a material effect on the consumer, for example, changes in the composition of the food or introduction of allergens. Several countries require labelling when the final product is different than the conventional food product, regardless of whether the difference has no consequences for health or the quality of the product. There is less agreement on whether final products which do not contain any GM material should be labelled if they were derived from a GM crop and whether a food should be labelled because of the process of production. A few countries explicitly address the use of negative food labels, that is, labels that claim that a food does not contain GM ingredients.

Of particular significance is the fact that a number of countries have set thresholds for the unintentional presence of GM material. Unintentional or adventitious presence can occur when pollen flows from GM crops to conventional or organic crops and when GM DNA comes into contact with other foods in farm equipment, storage silos, transport containers and food processing plants. Very minute quantities of GM DNA can lead to a positive test result for GM contents even though the food was produced through conventional or organic methods.

Table 10.2 Labelling requirements for genetically modified foods from different countries

Country	Main legislation	Main features of the policy		
		Positive label	Unintentional GM presence	Negative label
Argentina	Argentina Food Code Law 24.240 on Consumer's Defence Commercial Loyalty Law	'…specific regulations at a national level are based on the characteristics and properties of the product when these are technically verifiable…' (p. 2) No law at national level for labelling food produced from raw material or ingredients derived from genetically modified organism.		
Australia	Australia New Zealand Food Standards Code Standard 1.5.2– Food Produced Using Gene Technology Trade Practices Act, 1974	'GM foods and food ingredients [including food additives and processing aids] must be labelled if there is novel DNA and/or novel protein in the final food, or if the food has "altered characteristics"… [that is] significantly different from its non-GM counterpart with respect to allergenicity, toxicity, nutritional impact or end use.' (p. 7) '…does not require mandatory labelling on the basis of method of production where there is no novel DNA or novel protein.' (p. 7)	'… no more than 10 g/kg per ingredient … [is permitted to] remain unlabelled'.	'… might be called on to substantiate the claim…'
Brazil	Decree 4.680 of April 24, 2003 Portaria (Regulation) 2.658 December 22, 2003 Law 8.078 of September 11, 1990 Code of Defense of the Consumer Law 11.105 of March 24, 2005	'…labelling of foods and food ingredients containing or consisting of organisms obtained by certain techniques of genetic modification/genetic engineering is mandatory…' '…the main reason for the labelling … is to guarantee the legitimate consumer right to information, in order to favour his/her conscious choice of foods.' (p. 10)	Must inform about presence of GMO when above the limit of 1%.	

Canada	Food and Drug Regulations Food and Drugs Act Consumer Packaging and Labelling Act Competition Act National Standard for Voluntary Labelling and Advertising (Draft)	'… mandatory labelling requirements when there is a health and safety change or a significant change in nutrition or composition in the novel food (including products of genetic engineering), and voluntary labelling requirements for method of production labelling.' (p. 12) '… permit voluntary … positive labelling on the condition that the claim is not misleading or deceptive and the claim itself is factual' (p.12)		'…Permit voluntary … negative labelling on the condition that claim is not misleading or deceptive and and the claim is factual'
European Community	Article 2 of Directive 2000/13/EC of the European Parliament and of the Council of 20 March 2000 Regulation (EC) No. 1830/2003 and Regulation (EC) No. 1829/2003 of the European Parliament and of the Council of 22 September 2003	'…labelling … should include objective information … that a food consists of, contains or is produced from GMOs. Clear labelling, irrespective of the detectability of DNA or protein resulting from the genetic modification to the final product … facilitates informed choice and precludes potential misleading of consumers as regards methods of manufacture or production.' (p. 23) 'labelling should give information about any characteristic or property which renders a food different from its conventional counterpart with respect to composition, nutritional value or nutritional effects, intended use of the food and health implications for certain sections of the population, as well as … ethical or religious concerns' (p. 23)	'…a proportion no higher than 0.9% of the food ingredients considered individually or food consisting of a single ingredient…'	

Table 10.2 *continued*

Country	Main legislation	Main features of the policy		
		Positive label	Unintentional GM presence	Negative label
Ghana	National Biosafety Bill (draft)	No Ghana Standards or regulations (as of February 2007) (p. 33)		
India	Prevention of Food Adulteration Rules, 1955 37 E Labelling of Genetically Modified Food (draft)	'a GM Food, derived there from, whether it is primary or processed or any ingredient of food, food additives or any food product that may contain GM material shall be compulsorily labelled, without any exception.' (pp. 36–37) '…provisions will be applicable to all such products both imported or domestically produced' (pp. 36–37) 'the label of imported GM Food or derived there from … shall also indicate that the product has been cleared for marketing and use in the country of origin so that the verification, if needed can be taken up with that country without having to resort to testing.' (pp. 36–37)		
Japan	Article 21 of the Enforcement Regulation of the the Food Sanitation Law The Labeling Standard for Genetically Modified Foods (Notification No. 517 of the Ministry of Agriculture,	'…labelling is required for the products in which genetically modified DNA or protein is present and detectable.' (p. 39) 'Processed foods in which DNA or protein is undetectable are not subject to mandatory labelling…' Labelling is mandatory for 'GM foods whose composition or nutritional values are significantly different from their conventional counterparts.'	Adventitious presence accepted up to 5%.	'Non-GM products may be voluntarily labelled as "non-GM" if certification is provided to show that the non-GM ingredients were under the identity preserved handling…'

	Forestry and Fisheries of March 31, 2000 Law Concerning Standardization and Proper Labeling of Agriculture and Forestry Products	
Malaysia	Drafting regulations for mandatory labelling (p. 43)	
Mexico	Genetically Modified Organisms Biosafety Law Article 101 General Health Law Article 282 Bis 2 Statute for the Sanitary(safety) Control of Products and Services Article 166 Mexican Official Standards (Technical Regulations)	Labelling of GMOs and of products containing them is required: 'in the events where their traits are significantly different than those of conventional products … explicit reference must be made to "genetically modified organisms" and the label must state their food composition or such nutritional properties that are different from their conventional counterparts.' (p. 44) There is no obligation to label where the GMO is not different from its conventional counterpart. Labelling is not required solely because of the process or method of production.

Table 10.2 *cotinued*

Country	Main legislation	Main features of the policy		
		Positive label	Unintentional GM presence	Negative label
New Zealand	Australia New Zealand Food Standards Code Food Standard 1.5.2 'Food Produced using Gene Technology' Standard 1.2.9 'Legibility Requirements' Fair Trading Act of 1986	'…the Code requires all foods, food ingredients or additives sold … to be labelled at point of sale, where novel DNA or protein is present in the final food, or the food has altered characteristics as a result of genetic modification processes.' (p. 47) Flavourings making up less than 1% are exempt from labelling. 'The GM labelling requirements apply to all packaged and bulk foods, but do not apply to food prepared in restaurants, cafes and takeaways.' '…does not require mandatory labelling for method of production, where a food has been derived from gene technology, but does not contain novel DNA and/or novel protein.'	…trace amounts of GM material (less than 1%)…'	
Norway	Regulations relating to the labelling, transport, import and export of genetically modified organisms (GMOs) General Regulation of 8th July 1983 no 1252 Section 16a Regulation of 21st December	'the regulations contain rules for the authorisation, labelling and traceabilty of both GM food and feed. The regulations are based upon EU Regulations (EC) Nos 1829/2003 … and 1830/2003…' (p. 52) 'The labelling regulations apply to all GM foods including GMOs and food derived from GMOs, whether their properties or characteristics be different from those of comparable conventional food or not.'	Label required if GM component constitutes more than 0.9% of the ingredient.	

	1993 no. 1385 Section 4a no. 4 Section 7 and Section 10	
United States of America	Federal Food, Drug and Cosmetic Act (FFDCA) Section 403(a)(1) of the FFDCA (21 U.S.C. 343(a)(1) and Section 201(n) of FFDCA (21 U.S.C. 321(n) Food and Drug Administration Guidance for Industry: Voluntary Labeling Indicating Whether Foods Have or Have Not Been Developed Using Bioengineering: Draft Guidance 2001	'No special labelling requirement for bioengineered foods as a class of foods.' 'If a bioengineered food is significantly different from its traditional counterpart … the name must be changed to describe the difference.' 'If an issue exists … regarding how the food is used or consequences of its use, a statement must be made in the labelling to describe the issue.' 'If a bioengineered food has a significantly different nutritional property, its labelling must reflect the difference.' 'If a new food includes an allergen that consumers would not expect to be present … that allergen must be disclosed…' 'All statements … must be truthful and not misleading.' (p. 57)

Source: extracted from the Report of the CCFL Working Group on Labelling of Foods and Food Ingredients Obtained Through Certain Techniques of Genetic Modification/Genetic Engineering held in Oslo, 6–7 February 2007 (CAC, 2007).

Unintentional presence can undermine consumer confidence in the integrity of the food label.

10.2.2 Case study: voluntary labelling in the United States of America

In the USA, the Food and Drug Administration (FDA) has responsibility for regulating all processed and packaged foods, animal feed, food additives, veterinary drugs and human drugs that are derived from agricultural biotechnology (Executive Office of the President, Office Science and Technology Policy,1986). The authority to regulate labels for GM foods labels is derived from the Food, Drug and Cosmetic Act of 1938 which states that the labelling of a product must 'reveal facts material in the light of such representations or material with respect to consequences which may result from the use of the article' (United States Congress, 1938). In addition, it is illegal to misbrand a food through labelling which is 'false or misleading in any particular …' (*ibid*).

The agency's approach to regulation of GM foods was explained in 1992, when the FDA issued the 'Statement of Policy: Foods Derived from New Plant Varieties; Notice". The 1992 policy stated:

> The regulatory status of a food, irrespective of the method by which it is developed, is dependent upon objective characteristics of the food and the intended use of the food (or its components). Consumers must be informed, by appropriate labeling, if a food derived from a new plant variety differs from its traditional counterpart such that the common or usual name no longer applies to the new food, or if a safety or usage issue exists to which consumers must be alerted. (Food and Drug Administration, 1992, 22991)

In 2001, after reviewing its approach in light of public, industry and trade concerns, the FDA announced a draft policy: 'Guidance for Industry, Voluntary Labeling Indicating Whether Foods Have or Have Not Been Developed Using Bioengineering' (FDA guidance) (Food and Drug Administration, 2001). The main features of the guidance are shown in Table 10.3.

It is notable that the FDA approach provides guidance for producers who wish to inform consumers that their product does not contain GM ingredients, i.e. negative labelling. In addition, it specifically draws attention to the United States Department of Agriculture rules for organic foods (National Organic Program final rule; 65 FR 80548) involving requirements for certification that a product is organic. 'The national organic standards would provide for adequate segregation of the food throughout distribution to assure that non-organic foods do not become mixed with organic foods. The agency believes that the practices and record keeping that substantiate the "certified organic" statement would be sufficient to substantiate a claim that a food was not produced using bioengineering.' (Food and Drug Administration, 2001).

Table 10.3 Key features of the FDA, 2001 guidance for voluntary labelling of bioengineered foods

Bioengineered
Optional to say 'contains (product) developed/produced through biotechnology'
Allowed to claim 'developed through biotechnology because (positive reason)' but **must substantiate claim**. (emphasis added)
Cannot claim benefits for whole product if amount of positive ingredient insignificant
Must disclose allergens not found in conventional counterpart
Must change name if significantly different
Optional to say 'contains (product) developed/produced through biotechnology'
Allowed to claim 'developed through biotechnology because (positive reason)' but **must substantiate claim**. (emphasis added)
Cannot claim benefits for whole product if amount of positive ingredient insignificant
Must disclose allergens not found in conventional counterpart
Must change name if significantly different
Label may apply to human foods and animal feeds
Non-bioengineered
All ingredients must be non-bioengineered
Cannot imply that specific product is non-bioengineered if no products of this type are bioengineered.
Can say all foods of a type are non-bioengineered
Must be able to substantiate 'non-bioengineered' through testing, documentation, segregation
USDA certified organic foods are non-bioengineered by definition
Permitted to say biotechnology not used if there is **no suggestion that product is superior**. (emphasis added)
Label may apply to human food and animal feeds

Source: Adapted from FDA, 2001.

10.2.3 Case study: mandatory labelling in the European Union

In the European Union, the European Commission, the European Parliament and the European Food Safety Authority (EFSA) have responsibilities for developing laws to regulate GM products. Regulations in each EU country must be harmonized with the regulations of the other members of the EU so that foods can flow freely throughout the European market.

Since the late 1980s, the governments that now comprise the EU have considered genetically modified organisms as a distinct class of biological entities requiring special regulatory attention (Lezaun, 2006). This process-oriented approach has been influenced by the 'Precautionary Principle' which states 'where there are threats of serious or irreversible damage, lack of full scientific certainty shall not be used as a reason for postponing cost-effective measures to prevent environmental degradation' (Rafferty, 2004, 282). The precautionary principle stems from Principle 15 of the Rio Declaration on Environment and Development that has the following aims: ... '[T]o contribute to ensuring an adequate level of protection in the field of safe transfer, handling and use of living modified organisms resulting from modern biotechnology that may have adverse effects on

Table 10.4 Key features of the European Union's mandatory labelling law for genetically modified foods

Where the food consists of more than one ingredient, the words 'genetically modified' or 'produced from genetically modified (name of the ingredient)' shall appear in parentheses immediately following the ingredient or a footnote.
Where the ingredient is designated by the name of a category, the words 'contains genetically modified (name of organism)' or 'contains (name of ingredient) produced from genetically modified (name of organism)' shall appear in the list of ingredients or a footnote.
Where there is no list of ingredients, the words 'genetically modified' or 'produced from genetically modified (name of organism)' shall appear clearly on the labelling.
Where there is no list of ingredients, they shall appear clearly on the labelling.
Where the food is offered for sale to the final consumer as non-pre-packaged food, or as pre-packaged food in small containers, the information must be permanently and visibly displayed either on the food display or immediately next to it, or on the packaging material, in a font sufficiently large for it to be easily identified and read.
The law does not apply to foods containing GM material of less than 0.9 percent if the presence of the GM ingredient is adventitious or technically unavoidable. Lower thresholds may be established for particular foods or to take into account scientific and technical advances.

Source: Adapted from European Parliament, 2003a.

the conservation and sustainable use of biological diversity, taking also into account risks to human health ...' (Convention on Biological Diversity, 2000).

In 2003, the European Parliament enacted two complementary laws regarding GM food: Regulation (EC) 1829/2003 requires labelling for human food and animal feed containing genetically modified organisms, 'to enable consumers to make an informed choice,' while Regulation (EC) 1830/2003 'guarantees the traceability and labeling of genetically modified organisms and products produced from GMOs throughout the food chain ... to facilitate monitoring'(European Parliament, 2003a,b). The law requires that operators throughout the food chain keep records of their use of GM products and that this be declared on a food package if the content of GM material exceeds 0.9 percent. The main features of the regulations are shown in Table 10.4.

A consumer in Europe would assume that an unlabelled product does not contain GM ingredients because there is a mandatory positive label, i.e. those products that do contain such ingredients must be labelled.

The EU decided not to include provisions for negative labelling in their legislation because 'experiences in some Member States revealed that voluntary "GMO-free" (or similarly phrased) schemes were beset by a number of technical, commercial and other difficulties.' (CAC, 2008, 46).

10.3 Commercial experiences with labelling

Although some early GM products were labelled (Martineau, 2001), there has been very little published evidence regarding companies' implementation of labelling policies in recent years. Indeed, when CCFL members were asked about experiences, a number of countries reported that products had been tested yet almost none had been found to have GM material in sufficient quantity as to require labelling. Therefore, there was no recent practical experience with positive labels that is known to governments.

There may be several explanations for this lack of information, such as a lapse between the times that a policy is enacted and when the labelled products reach market shelves and the time and resources that are required for monitoring and documenting industry and consumer reactions to labelled products. However, given the keen interest in this topic, it seems that there would have been efforts made to monitor labelling experiences. A more likely reason for the lack of experience in implementing the labelling policies is the lack of interest among food producers and retailers in selling foods that are labelled as containing GM ingredients. In the case of labelling foods as 'non-GM' there is also reluctance since the regulations are perceived to be burdensome.

If a farmer or manufacturer wished to sell GM foods, food retailers in some markets such as Europe act as 'gatekeepers' and prevent these foods from being available because of their scepticism about consumer acceptance of GM foods (Knight *et al.*, 2005). This experience was expressed by the European Community delegation to CCFL; they reported that

> few food products labelled as genetically modified are at the present time on the Community market. The situation is however not uniform throughout the EU since in some Member States the number of GM products is negligible while in others their number is more significant … The sale of this type of products is mainly governed by factors that are not related to the legislative framework, such as consumer demand and the policies of food producers and retailers. (CAC, 2007, 28).

10.3.1 Disincentives to label food products

It is generally acknowledged that the generation of GM crops that are currently cultivated have agronomic traits that appeal to many farmers; they do not have qualities that might attract consumers. On the contrary, in research in the large and affluent markets of North America and Europe study participants express their preferences for foods that are not produced with GM ingredients (Evenson and Santaniello, 2004; The Mellman Group, Inc., 2006). Given the present milieu, there is little incentive for the food industry to use positive labels, i.e. statements that claim that a food does contain GM ingredients, while there may be some incentive to use negative labelling, i.e. the claim that a food does not contain GM ingredients.

10.4 Conclusions

If food producers, manufacturers and retailers are wary of consumer reactions to foods that are labelled as GM, they will not implement a labelling policy, whether it is mandatory or voluntary. Food sellers volunteer to label a product when they believe it will encourage sales. When it is mandatory to disclose information about a food that may deter consumers from buying the product, food sellers avoid the risk of labelling. In the case of GM foods, they may reformulate their products and sell conventional and organic products. In the case of negative labels, producers may be deterred from labelling because the costs of substantiating this claim may not be justified by the premiums consumers are willing to pay and the risks that a label may be considered to be misleading and in violation of regulations.

Regardless of how well-intentioned and well-designed a policy may be, it appears that there is little implementation of labelling policies when it comes to GM foods. Without substantial experience and evidence to demonstrate the feasibility and usefulness (or lack thereof) of a specific approach, it will be difficult for governments to move forward to reach consensus on a harmonized standard or guideline for labelling of GM foods. The CCFL delegates will continue to discuss recommendations for the labelling of foods and food ingredients obtained through genetic modification (CAC, 2009). For the foreseeable future, each country will develop its own policies, in keeping with its own priorities, as well as interpretation of the existing Codex standards and international agreements.

10.5 References

CAC (2007) Codex Committee On Food Labelling, Thirty-Fifth Session, Ottawa, Canada, 30 April–4 May 2007, Labelling Of Foods And Food Ingredients Obtained Through Certain Techniques Of Genetic Modification/Genetic Engineering: Proposed Draft Guidelines For The Labelling Of Foods And Food Ingredients Obtained Through Certain Techniques Of Genetic Modification/Genetic Engineering : Labelling Provisions (At Step 4). Report of The CCFL Working Group On Labelling of Foods And Food Ingredients Obtained Through Certain Techniques Of Genetic Modification/Genetic Engineering, Oslo 6–7 February 2007. Joint FAO/WHO Food Standards Programme. Rome: FAO/WHO.

CAC (2008) Codex Committee On Food Labelling, Thirty-Sixth Session, Ottawa, Canada, 28 April–2 May, 2008, Labelling of Foods and Food Ingredients Obtained Through Certain Techniques Of Genetic Modification/Genetic Engineering: Proposed Draft Guidelines for the Labelling of Foods and Food Ingredient Obtained Through Certain Techniques Of Genetic Modification/Genetic Engineering: Labelling Provisions (At Step 4). Report of The Working Group on the Labelling of Foods and Food Ingredients Obtained Through Certain Techniques of Genetic Modification / Genetic Engineering, Accra, 28–30 January 2008. Joint FAO/WHO Food Standards Programme. Rome: FAO/WHO.

CAC (2009) Codex Alimentarius Commission Thirty-second Session, Rome, Italy, 29 June–4 July 2009. Report of the Thirty-seventh Session of the Codex Committee on Food Labelling, Calgary, Canada, 4–8 May 2009. Joint FAO/WHO Food Standards Programme. Rome: FAO/WHO.

CONVENTION ON BIOLOGICAL DIVERSITY (2000) Cartagena Protocol on Biosafety to the Convention on Biological Diversity. January 29. www.cbd.int/doc/legal/cartagena-protocol-en.doc. Retrieved May 27, 2007.

EUROPEAN PARLIAMENT (2003a) Regulation (EC) No 1829/2003 of the European Parliament and of the Council of 22 September 2003 on genetically modified food and feed. *Official Journal of the European Union* L268: 1–23.

EUROPEAN PARLIAMENT (2003b) REGULATION (EC) No 1830/2003 of the European Parliament and of the Council of 22 September 2003 concerning the traceability and labelling of genetically modified organisms and the traceability of food and feed products produced from genetically modified organisms and amending Directive 2001/18/EC. *Official Journal of the European Union* L268/24: 24–28.

EVENSON, R.E. AND SANTANIELLO, V. (Editors) (2004) *Consumer Acceptance of Genetically Modified Foods*. Oxford: CABI Publishing.

EXECUTIVE OFFICE OF THE PRESIDENT, OFFICE OF SCIENCE AND TECHNOLOGY POLICY (1986) 'Coordinated Framework for Regulation of Biotechnology' Action: Announcement of policy; notice for public comments. *Federal Register* 51, 23302, June 26.

FOOD AND DRUG ADMINISTRATION (1992) Part IX Department of Health and Human Services, FDA, 'Statement of Policy: Foods Derived from New Plant Varieties; Notice'. *Federal Register*, 57, No. 104, 22984–23001, May 29.

FOOD AND DRUG ADMINISTRATION (2001) 'Draft Guidance for Industry: Voluntary Labeling Indicating Whether Foods Have or Have Not Been Developed Using Bioengineering; Notice'. *Federal Register*, 66, No. 12, 4839–4842, January 18.

JAHN, G., SCHRAMM, M. AND SPILLER, A. (2005) 'The reliability of Certification: Quality Labels as a Consumer Policy Tool'. *The Journal of Consumer Policy* 28: 53–73.

JAMES, C. (2008) *Global Status of Commercialized Biotech/GM Crops: 2008 ISAAA Brief 39*. Ithaca, NY: International Service for the Acquisition of Agri-Biotech Applications.

KNIGHT, J.G., MATHER, D.W. AND HOLDSWORTH, D.K. (2005) 'Impact of genetic modification on country image of imported food products in European markets: Perceptions of channel members'. *Food Policy* 30: 385–398.

LEZAUN, J. (2006) 'Creating a New Object of Government: Making Genetically Modified Organisms Traceable'. *Social Studies of Science* 36, no. 4 (August): 499–531.

MARTINEAU, B. (2001) *First Fruit, the Creation of the Flavr Savr™ tomato and the Birth of Biotech Food*. New York: McGraw-Hill.

THE MELLMAN GROUP, INC. (2006) Review of public opinion and research. Philadelphia, PA: Pew Initiative on Food and Biotechnology. Available at: http://www.pewtrusts.org/uploadedFiles/wwwpewtrustsorg/Public_Opinion/Food_and_Biotechnology/2006summary.pdf [accessed July 2009].

RAFFERTY, B. (2004) 'The Door Opens Slightly: Recent European Union Regulations on Genetically Modified Products and the Ongoing United States-European Union GM Product Dispute'. *Georgetown International Environmental Law Review* 16, no. 2 (Winter): 281–300.

UNITED STATES CONGRESS (1938) Federal Food, Drug and Cosmetic Act, Washington, DC.

Index